Dusty Earthlings

Dusty Earthlings

*Living as Eco-Physical Beings
in God's Eco-Physical World*

John Mustol

Foreword by
Nancey C. Murphy

The Lutterworth Press

For Betsy

The Lutterworth Press
P.O. Box 60
Cambridge
CB1 2NT
United Kingdom

www.lutterworth.com
publishing@lutterworth.com

ISBN: 978 0 7188 9523 5

British Library Cataloguing in Publication Data
A record is available from the British Library

First published by The Lutterworth Press, 2017

Copyright © John Mustol, 2012

Published by arrangement
with Cascade Books

The Holy Bible, New International Version®, NIV Copyright © 1973, 1978, 1984, 2011 by Biblica, Inc.™ Used by permission. All rights reserved worldwide.

All rights reserved. No part of this edition may be reproduced, stored electronically or in any retrieval system, or transmitted in any form or by any means, electronic, mechanical, photocopying, recording, or otherwise, without prior written permission from the Publisher (permissions@lutterworth.com).

Contents

Foreword by Nancey C. Murphy / ix
Preface / xiii
Acknowledgments / xvii

1. Introduction: Humans as Dusty Earthlings / 1
2. Our Separation from Creation / 30
3. Human Nature according to Science / 54
4. The Bible I: God and Humans / 76
5. The Bible II: The Image of God,
    the Dominion Mandate, and Other Issues / 118
6. Ecology for Dusty Earthlings / 154
7. Ethics for Dusty Earthlings:
    Living in God's Eco-Physical World / 202
8. Conclusion / 227

Glossary / 251
Bibliography / 259
General Index / 274

Foreword

THIS WORLD IS OUR *home!* John Mustol has provided here one of the most powerful theological and scientific arguments for this conclusion. He laments the fact that we Christians perceive ourselves as separate from the rest of God's creation, particularly from nature. The root cause of this, as of other forms of alienation, is sin, but, he says, "it has been exacerbated by philosophical, religious, cultural, scientific, and technological developments in the modern era."[1]

In this foreword, I'd like first to suggest a very significant philosophical development (but one that goes all the way back to early Christian apologists) that has been a major cause of our alienation. Then I'll turn to developments in both science and biblical studies that are calling us "back home."

Diogenes Allen, in his lovely book *Philosophy for Understanding Theology*, subtitles a chapter on Plato's philosophy, "This World Is Not Our Home."[2] Allen points out that when early theologians encountered Plato's creation story in the dialogue *Timaeus*, they thought that either Plato had somehow read Genesis or he had received his knowledge by divine revelation.[3] Because Plato's works were so useful in addressing the Christian message to pagan audiences, a great deal of his thought became incorporated into the Christian tradition. However, throughout the twentieth century, biblical scholars and theologians have concluded that much of his philosophy had displaced earlier, more authentic Christian teachings.

I have become famous (or infamous, depending on one's point of view) for promoting a physicalist account of human nature; I, along with many others, believe that body-soul dualism was a later addition

---

1. See p. 227 of this volume.
2. Allen, *Philosophy*, 39.
3. Ibid., 15.

to Christian thought, not found in the Bible, and that it has had pernicious effects, only one of which is the tendency of so many Christians to see ecological issues as irrelevant to Christian ethics. Physicalism and dualism are each bound up with an entire worldview. Dualism belongs to a worldview that owes a great deal to Plato. He invented the notion of a nonmaterial realm transcending this corruptible material world. The dualist view of the person mirrored this cosmic dualism. The human soul, immortal, belongs to the transcendent realm of the Forms (or Ideas), and life in the body is temporary imprisonment. Value resides in the other world; in fact, some of Plato's followers counted matter as essentially evil. The Western imagination has been deeply influenced by this otherworldliness, and Christians have focused more on a Platonic hope for the soul's escape, to live forever in a transcendent heaven, than what has come to be seen by scholars as a more accurate reflection of the gospel.

By the middle of the twentieth century many biblical scholars and theologians had come to see body-soul dualism as an import from Greek and Roman philosophy. More importantly, they recognized that the good news Jesus preached was not about getting to heaven, but rather about the kingdom of God, already "come near" in his person (Mark 1:15). Evangelical scholar George Eldon Ladd is now famous for his insistence on the centrality of the kingdom of God to New Testament teaching. This is not, of course, to deny the afterlife. It is rather to emphasize the importance of bodily resurrection. If the heart of Jesus' preaching was the kingdom of God, then we might say that the heart of Paul's was the bodily resurrection of Jesus, and its implications for our own future.

Looking forward to the resurrection and transformation of our bodies leads naturally to the expectation that the entire cosmos will be similarly transformed. In Jesus' resurrection we see the first fruits of the transformation for which the whole creation is longing. As the Apostle Paul says: "The creation waits in eager expectation for the sons of God to be revealed. For the creation was subjected to frustration, not by its own choice, but by the will of the one who subjected it, in hope that the creation itself will be liberated from its bondage to decay and brought into the glorious freedom of the children of God" (Rom 8:19–21).

I turn now from biblical studies to science. Science, too, provides grounds for the recognition that this world is indeed our home. Mustol rightly emphasizes the significance of the story of our creation out of the dust of the ground. What cosmology now tells us is that this dust is ultimately star dust. The heavy elements form in stars and are distributed when the stars explode—to become planets and plants and people.

In cosmology there is a fascinating discussion of the so-called anthropic issue. This is based on calculations showing that very, *very* small changes in any of the numbers that go into the basic laws of physics would have resulted in a universe in which no life is possible. For instance, if the strength of gravity had been slightly higher, our universe would have collapsed in on itself too quickly for stars, planets, and life to evolve. These anthropic calculations have led some interpreters to describe the universe as fine-tuned for life. Some see it as the work of God; others dismiss it as mere chance. But it has led a number of thoughtful scientists to raise questions about the significance of human life in the universe. "I do not feel like an alien in the universe," says physicist Freeman Dyson.[4] Indeed, this world is our home. The vastness of the universe was once taken to speak to human insignificance. But in light of the anthropic calculations, we can say that the universe needed to be as immense as it is, and as old as it is, and as full of stars as it is, in order for us to be here.

If, in light of current biblical research, current developments in cosmology, and (most important for readers of this book) current developments in the science of ecology, we reject the Platonic vision of the "flight of the alone to the Alone" and return to the biblical view of the rule of God "on earth as it is in heaven," we find a vision of the end of time that shows the ultimate value of history; that shows that history is meaningful, for past achievements are not left behind but transformed, and past sorrows add poignancy to present joy. Finally, it is a vision that shows there to be ultimate value in our care for and harmony with the whole of nature. This will be a world whose character Isaiah evoked in his prophecy:

> "Behold, I will create new heavens and a new earth. . . .
> I will create Jerusalem to be a delight and its people a joy. . . .
> The sound of weeping and crying will be heard in it no more. . . .
> They will build houses and dwell in them;
> they will plant vineyards and eat their fruit. . . .
> Before they call I will answer;

---

4. Dyson, *Disturbing the Universe*, 250.

Foreword

> while they are still speaking I will hear.
> The wolf and the lamb will feed together, and the lion shall eat straw like the ox . . . .
> They will neither harm nor destroy in all my holy mountain," says the Lord.
> Isa 65:17–25, *passim*

Note that this is a social *and* ecological vision—the re-creation of earthly life. It is a vision of unimpaired, immediate relation to God. It is a vision of a whole new cosmos—new heavens as well as new earth—in which humankind *and all of nature* will be reconciled.

Nancey C. Murphy
Professor of Christian Philosophy
Fuller Theological Seminary

Preface

WE CHRISTIANS HAVE STRUGGLED with "physicalness" and with physical existence for centuries.[5] We seem to be perpetually conflicted about our bodies, about the physical world around us, and about the sciences that study our bodies and the physical world. But in truth, Christianity ought to be affirming of physical existence. We are the ones, after all, who worship the God who created it all and who redeemed it by becoming physical flesh himself and living on earth as a human (John 1:14).[6] God literally became a physical being who required food, grew, learned, talked, walked, metabolized nutrients, eliminated waste, felt pain, bled, and depended for life on the Palestinian ecosystem in which he lived. Jesus, like all of us, was a *dusty earthling*—bone and blood amidst the dirt and rocks of planet earth. Even in his post-resurrection state, Scripture says, he was "flesh and bones," a physical being (Luke 24:39). We Christians ought to stop being so ambivalent about our bodies and the physical world and embrace the earthy, organic character of our theology and ourselves as part of God's earthly creation. Today in the early twenty-first century, we humans are facing an array of ecological problems stemming from our troubled relationship with the rest of the physical creation. These problems will almost certainly increase as this century unfolds. We Christians need to engage with ecology. Herein is a small contribution to that engagement.

In this book I argue that we are thoroughly and completely physical beings—part and parcel of God's good creation. And as physical beings, we are also ecological beings—what I call *eco-physical beings*—drawing our life and sustenance from the ecosystems that surround and support us along with all creatures on God's good earth. As eco-physical beings

---

5. Peterson, "In and of the World?" 242.
6. I have elected to use conventional male pronouns for God, but clearly God is not gendered—neither male nor female. After all, he created gender. I hope this does not offend my sisters (or brothers) in Christ who read this book. In all other contexts I have tried to use gender-neutral language.

we are subject to all the principles, patterns, parameters, and limits (ecological realities) that govern all of life on this planet. Moreover, as the uniquely ethical species made in the image of God, capable of genuine moral life and a special relationship with God through Jesus Christ, we are called to embodied eco-physical living and nurturing oversight of God's earthly creation, all to his glory. This is the message of this book.

In chapter 1, I introduce the basic argument of the book and discuss what I call the *Ecological Problem*—the array of difficulties we face in the world today stemming from our troubled relationship with the rest of creation. I make a claim for the thoroughly physical nature of Christian spirituality and its relationship to the Ecological Problem, which is, at its origin, spiritual in nature.

In chapter 2, I speak to the premise that we are separated from the rest of creation. Our broken relationship with the rest of creation arises in a primal sense from our rebellion and sin, but it has been exacerbated by historical, philosophical, cultural, and religious factors in Western culture, especially in the modern period. I survey these and show how they have contributed to where we are today.

Chapter 3 focuses on scientific findings that relate to our human nature. Modern neuroscience and biology are showing more and more our commonalities with other animals. I examine some of the scientific developments that affirm our physicality, our kinship with other creatures, and our integral membership in God's eco-physical world. At the same time, I review scientific findings that affirm our uniqueness among God's creatures and that justify God's placing us over his creation (Gen 1:26–28). We are the *ethical species*—the one species on earth capable of managing God's world such that God can hold us responsible for it. We are dusty earthlings to be sure, but we are special dusty earthlings.

In chapters 4 and 5, I look into Scripture for the theological principles that pertain to my argument. Chapter 4 focuses on God and humans—how a biblical grasp of the nature of God and of ourselves helps us understand our place within the world. Chapter 5 examines some additional relevant biblical issues including the image of God, the dominion mandate (Gen 1:26–28), ecological sin, theocentrism, and God's redemption of creation.

In chapter 6, I describe some of the principles, patterns, parameters, and limits that govern our earthly eco-physical existence. This includes feedback loops, nutrient cycles, energy flow and exchange, food

Preface

webs, population dynamics, and so on. In the course of this discussion, I note some of the implications of these ecological realities for us dusty earthlings living on God's earth.

In chapter 7, I discuss ethics. As dusty earthlings called to follow Jesus, how should we live and what should we do? I offer two virtues—humility and self-control—and two principles—kenosis and justice. And I argue that we should rejoin creation in praising God and suggest some ways to do that.

In chapter 8, after summarizing, I conclude with some comments on topics relevant to my thesis including what I call "eco-consciousness," ecological education, individual action and public policy, population, economics, and a look to the future.

My goal is to take Jesus, Scripture, our physicality, and God's physical creation seriously and try to see how these relate to the ecological problems of our day. I offer answers to some questions, leave some questions unanswered, and raise still others. My goal is not to be exhaustive or to make final statements but to move the conversation along. If you join with me in this endeavor, maybe both of us will come a little closer to following Jesus in a more authentic way in his world.

I am an evangelical, and I have tried to write the book from that perspective. I hope my fellow evangelicals will read it—especially conservatives. While they have been rightly concerned with the primacy of Scripture and the importance of orthodoxy, they have been reluctant to engage with ecology. Their reasons for this are understandable, but ecological problems are here to stay. It's time for *all* of us Christians to engage the problem and bring our prodigious biblical and theological resources to bear. For Christians, ecology and creation care are integral to our faith and life in Christ within his creation. (They are not merely political.) There are marvelous opportunities for ministry and witness to the gospel, in word and deed. I urge my conservative evangelical brothers and sisters to engage with ecology. I hope this book will encourage them to do so.

The book is my own, and I take full responsibility for it. I apologize in advance for errors. Without a doubt there are some. I welcome correction, comment, and response. For us dusty earthlings, life is learning, and I hope to learn much from this endeavor. My e-mail address is below; write to me and tell me what you think.[7]

---

7. john.mustol@gmail.com.

Acknowledgments

I would like to thank all who have helped me with this book. My son David offered valuable insights in many areas. My good friends Frank Kacer and theologian Dr. John Carstenson provided helpful criticism. Dr. Glen Scorgie, professor of theology and ethics at Bethel Seminary San Diego, has been a wonderful mentor and friend. He offered helpful guidance in the early stages when the book was little more than an idea. It is he who suggested the title Dusty Earthlings. I also offer my heartfelt thanks to Dr. Nancey Murphy, professor of Christian philosophy at Fuller Seminary for her helpful critique and ideas and her constant support. The seed for the book was a thesis I wrote for her. If not for her encouragement, this book would have never come to be. I also acknowledge the guidance of several great scholars and thinkers who do not know me, but I know them. Or, at least I know their writings. These include Joseph Sittler, Richard Bauckham, Malcolm Jeeves, R. J. Berry, Larry Rasmussen, Douglas John Hall, Steven Bouma-Prediger, Calvin DeWitt, Fred Van Dyke, Loren Wilkinson, Jürgen Moltmann, H. Paul Santmire, Wendell Berry, Ernst Conradie, James Nash, Holmes Rolston, and many others. I am thankful to all of them for their help and guidance. I also want to thank Joel Green, Warren Brown, and Glen Stassen for their teaching and guidance. My deepest gratitude goes to my wife Betsy who for thirty-eight years has always believed in me even when I didn't believe in myself. Her unwavering integrity, courage, and grace have always been a source of strength. I dedicate this book to her. Finally, thanks be to God and to the Lord Jesus Christ—Creator, Provider, and Redeemer of this beautiful creation of which we are a part. To him be the glory always.

1

Introduction: Humans as Dusty Earthlings[1]

"REMEMBER THAT YOU ARE DUST, and to dust you shall return."[2] So says the Catholic or Anglican priest while making crosses in black ash on the foreheads of kneeling parishioners. Called "the imposition of ashes," this ceremony is performed on Ash Wednesday each year in liturgical churches around the world. It marks the beginning of Lent, a forty-day period of prayer and repentance that ends with the celebration of Christ's resurrection at Easter. This declaration that we are dust recalls several passages of Scripture, all of which remind us of our humble state as part of God's earthly creation.[3] This is a vital truth. We humans *are* dust—physical beings, made of flesh and blood, arising from the soil of the earth and returning to it when we die. As these passages teach, and as this ceremony reminds us, *dustiness* or *physicality* is a basic fact of our human existence in this world. God has made it so.

I worked as a physician for over twenty years, and in that capacity I had many opportunities to observe my fellow humans in a variety of physical and spiritual conditions—in health and disease, in good times and bad. These experiences led me to question the traditional understanding I had been taught as a Christian, namely that humans are made up of two separate or separable entities: a material *body* and an immaterial or spiritual *soul*. (This understanding of human makeup is called body-soul dualism.) Again and again, I saw that physical things like chemicals (alcohol, drugs), broken bones, cancer, and pain could profoundly alter the spiritual life of people, and that spiritual things like

---

1. Throughout the book, please refer to the glossary for any word definitions you may need.
2. *Book of Common Prayer*, 265.
3. Gen 2:7, 3:19, 18:27; Pss 90:3, 103:14, 104:29; Eccl 3:20; 1 Cor 15:47.

faith, courage, hope, love, guilt, ignorance, loneliness, forgiveness, and prayer could deeply affect the physical life. It seemed to me that if these two parts of us did exist, they were very closely bound together.

I also realized that the only way we humans can be or do anything in the world is *physically*—in our bodies. We experience the world only through our senses of sight, hearing, touch, taste, and smell, and the only way we can do anything in the world, including communicate, is through the use of muscles—of the face for facial expressions, our mouths, throats, and chests for speech, our fingers for writing, and our arms and legs for working, walking, or whatever. Without muscles, perhaps we could think, we could believe, we could pray, we could worship, we could have spiritual experiences, but we couldn't *do* anything with our thoughts, our belief, our worship, or our experiences, and no one else would know about them. So I surmised that thinking, believing, worshiping, and even spiritual experiences are *physical* in that they require a reasonably intact and functioning brain and body in order to happen in any meaningful way. I concluded that in this life, on this earth, our bodies and souls are so tightly bound together that they form *one whole person*. When we speak of "spiritual" and "physical" or "soul" and "body," we are speaking not of separate entities or things but of different aspects or qualities of one, unified entity or thing—us, me and you. A unified, holistic understanding of human makeup, rather than a dualistic (body-soul) understanding, made a lot of sense.

On this point, it turns out that in recent decades, modern neuroscience—the study of the human brain and how it works—is discovering that human activities we traditionally attributed to the soul are, in fact, dependent on physical events in our brains and bodies.[4] Prayer, worship, decision-making, creativity, morality, and spiritual experience all appear to depend on physical events and processes in our brains and bodies. We humans seem to operate as "brain-bodies." Philosopher Thomas Nagel sums it up: "So far as we can tell, our mental lives and those of other creatures, including subjective experiences, are strongly connected with and perhaps strictly dependent on physical events in our brains and on the physical interaction of our bodies with the rest of the physical world."[5]

---

4. Murphy, "Science and Society," 126, 130.

5. Nagel, "Science and the Mind-Body Problem," in *What Is Our Real Knowledge about the Human Being*, quoted in Jeeves and Brown, *Neuroscience, Psychology, and*

Introduction: Humans as Dusty Earthlings

Also, biologists who study other animals are finding that we humans are not as different from them as we may have thought. Not only do we share DNA, proteins, metabolism, and a host of other chemicals and processes, we also share reasoning, self-awareness, future planning, choice, and social life. These things seem to occur, albeit in very simple or analogous ways, in nonhuman animals, affirming the physical nature that we share with them. In short, comparison with other animals shows that we too are animals—physical beings just like them.

I have a background in biology and have always enjoyed God's creation and his creatures. So ten years ago, when I began formal study of theology, I became interested in the theology of nature and how Christians approach the ecological problems of our day—how we are treating God's creation and his creatures. As I studied Scripture, I was impressed with the relevance and power of Christianity for ecology.[6] I was perplexed, however, to find Christians more or less oblivious. This made no sense. It seemed to me that we, of all people, who worship the creator and redeemer of all things, ought to be concerned about God's earth and his creatures. So why are Christians not engaged?

I believe that a major reason for this is that we misunderstand who and what we are as humans and how we fit into God's world. We Christians tend to think that our spiritual "soul" life is more or less separate from our physical "body" life and that it is far more important. Similarly, we have tended to see the physical world as *outside of* or *peripheral to* our "real" lives that are confined almost exclusively to our relationship with God and other humans. In our minds and hearts we are separated (or alienated) from the rest of God's creation. Although in reality we are an integral part of God's creation, in our perceptions and practices, we tend to see the physical, ecological world as "out there," outside of the realm of primary concern, and ourselves as "in here," inside the realm of primary concern—our personal, social, and economic life. I suggest that a deeper grasp of our physicality and our involvement in and dependence on the physical world will help us better understand who and what we really are as human beings and will show us that

---

Religion, 131.

6. Throughout the book, I will use the words *ecology* or *ecological* where others would use the words *environment* or *environmental*. These better express, I think, the holistic nature of the ecosystems in which we live and our integral membership within them. See "ecology, ecological" in glossary.

the physical-ecological world is important—vital, in fact, for our well-being and for our quest to live as God would have us live as his people in his world.

The Argument of the Book

So what am I trying to say in this book? My argument has four parts: (1) We are physical beings. (2) As physical beings we are embedded in and dependent on the physical world. (3) As embedded, dependent members of God's physical world, we are subject to all the ecological principles, patterns, parameters, and limits that define and circumscribe the existence of all God's creatures on his earth. (4) These realities have implications for our Christian ethics—how we should live in God's world. Let me unpack each of these in turn.

Part one says that we are physical beings. I have already touched on this, and in later chapters I shall present scriptural and scientific evidence for our physicality. Part two of my argument says that since we are physical beings, we are unavoidably embedded in and dependent upon God's physical, ecological creation. We cannot live or thrive on God's earth without the support and nourishment that its ecosystems provide. We are dependent, literally from moment to moment, on the planet's ecology. When we breathe, for example, we breathe out carbon dioxide that is absorbed by plants, and we breathe in oxygen that is entirely produced by those same plants. All that makes up our flesh—our muscles, bones, skin, brains, hearts—everything comes from and returns to the earth. All the food we eat comes from the earth, and all our waste, bodily and otherwise, must be absorbed and recycled by the earth's ecosystems. All the materials in our cars, clothes, buildings, computers, books, churches—everything comes from and returns to the earth. By necessity we *live within* and *depend upon* God's vast, earthly ecosystem—what we call the *ecosphere*. We humans form an integral part of this ecosphere. We are not outside of the ecosystem, and the ecosystem is not outside of our lives. We are unavoidably connected to other living and nonliving things—microbes, plants, animals, soil, rocks, wind, water, clouds, and all the natural cycles and processes that support and define life on God's good earth.[7]

---

7. Palmer, "Stewardship," 70.

This is summed up in the term *eco-physical*, a combination of the words *physical* and *ecological*. This neologism expresses, I believe, our basic nature as human beings, as *dusty earthlings*, denizens of this wondrous planet. We are physical beings, and as such we are ecological beings because, being *physical*, we must live within an ecosystem that supports and nurtures us. Our personal, social, economic, and, yes, our spiritual lives depend on our physical life and can only take place within the ecosystems of God's good earth. As theologian Larry Rasmussen puts it, "We live *in* earth *as* earth."[8] *Earth is our home.*[9]

Part three of my argument says that, since we are eco-physical beings, we are subject to all the principles, patterns, parameters, and limits of earthly existence—what I call *ecological realities*. In order for us to flourish (Gen 1:28), and in order for all God's creatures to flourish with us (Gen 1:22), we must obey these ecological realities. The science of ecology studies the relationships and interactions of living things with each other and with their environments.[10] It has developed a body of knowledge about these relationships and interactions including a set of principles, patterns, parameters, and limits (ecological realities) that apply to us humans just as they do to all creatures. We are not on the sidelines; we are *in* the game, so we have to play by the rules. Principles such as population growth patterns, resource limits, the carbon cycle, the water cycle, energy pyramids, and the law of entropy[11] govern our existence on this planet just as they do all living things.

The fourth and last part of my argument says that our eco-physical nature has implications for the kind of people we ought to be and how we ought to live in God's world. Knowing that we are eco-physical beings, embedded in God's eco-physical world, subject to its principles, patterns, parameters, and limits, helps us see how to live rightly and glorify God in our lives and in his world.

In summary, we are: (1) *physical beings*, therefore (2) *ecological beings*, therefore (3) we are subject to *ecological realities*, and finally (4) because of these things, we ought to be certain kinds of people as we live *in* God's earth.

8. Rasmussen, *Earth Community*, 276.
9. Jung, *We Are Home*, 69.
10. Molles, *Ecology*, 2.
11. This is a law of physics that is relevant to ecology. It is also called the second law of thermodynamics. I will discuss this in chapter 6.

Our Separation from Nature

At this point, you may be saying, "Of course we are physical beings, and we are part of the earth's ecosystem. So why write a book about it?" Well, our eco-physical nature may be self-evident, but we don't live as though it were so. Theologian Douglas John Hall says that the most important cause of ecological problems today is "humanity's seeming inability to understand its essential solidarity with all other forms of life, and to act upon that understanding."[12] Our economic system ignores the ecosphere within which it operates and on which it depends. Those of us who live in so-called developed societies such as the United States are living in a way that is overstressing the earth's capacity to support us. We live almost all of our lives within our constructed human environments of electronics, plastic, steel, glass, and concrete, isolating ourselves from God's natural world. We have used our wealth and technology to exclude nature from our lives and from our consciousness and so have lost touch with it. I recently saw a sign in the back window of a car with a large cross. Under it was the phrase, "Not of this world." Indeed, this is how we live—as though we were "not of this world"—not eco-physical beings, not part of the ecosystem, and not subject to the ecological realities that define earthly existence as God has designed it. Douglas Hall again writes: "There is a growing consensus among all who consider the future of life on earth that unless humankind achieves some profound awareness of its dependence upon nature, the future will be bleak—if it *is* at all. Most of those who have arrived at that conclusion, from whatever process of investigation and reflection, would add that such an awareness entails the clear recognition of our full human participation in nature, that we are not supranatural, after all, but part of the very process that we are gradually destroying."[13] In other words, we need what Hall calls a "radical conversion to creaturehood."[14] We must "re-enter God's creation."[15] This is what I am seeking to do in this book—to call us Christians back to our creaturehood, as God's

---

12. Hall, *Professing the Faith*, 335.
13. Ibid.
14. Ibid., 341.
15. Edward P. Echlin, "Let's Re-enter God's Creation Now," quoted in Bauckham, *Living with Other Creatures*, 144.

children, to reenter our "God-given home," as the great ancient church father Irenaeus called the created world.[16]

The Ecological Problem

Today, in the early twenty-first century, we face an array of ecological problems including population growth, overconsumption, maldistribution of resources, deforestation, land degradation, species loss, pollution, climate change, and so on. Some of these problems are controversial, but no one can deny that our relationship with God's earthly creation is deeply troubled and that we Christians need to address it.

For decades now, environmentalists have used the term *environmental crisis* to refer to these problems, but I think the word *crisis* should be dropped. Historian J. R. McNeill correctly remarks, "It is impossible to know whether humankind has entered a genuine ecological crisis. It is clear enough that our current ways are ecologically unsustainable, but we cannot know for how long we may yet sustain them, or what might happen if we do . . . The future, even the fairly near future, is not merely unknowable; it is inherently uncertain. Some scenarios are more likely than others, no doubt, but nothing is fixed."[17] As McNeill says, our current way of life is not sustainable, but we cannot know how that non-sustainability will be ended—whether by catastrophe, by government fiat, or by thoughtful, deliberate action. Furthermore, our problem is not about a crisis per se; it is about how we understand our mode of existence, our way of thinking and living, our vision of ourselves, of the world, and of God.[18] So instead of *crisis*, I prefer the term *Ecological Problem* coined by Francis Schaeffer in his 1970 book *Pollution and the Death of Man*.[19] This is not to deny that we are facing a crisis. We are, although, as McNeill says, it is a slow, irregular, fluctuating, complex, and unpredictable one. But the "crisis" language of environmentalists has been in use now for forty years, and it is worn out. Furthermore, the word *crisis* suggests a climax after which, once it has passed, we can relax and return to "normal life." This will never happen, and, as we

---

16. Santmire, *Travail of Nature*, 35.
17. McNeill, *Something New Under the Sun*, 358.
18. Hauerwas, *Vision and Virtue*, 36.
19. Schaeffer, *Pollution and the Death of Man*, 17.

shall see, it misunderstands the ecology of the planet.[20] The Ecological Problem will be with us from now on, and the changes we need to make to address it will themselves always be changing. We will never be able to relax and return to "normal life." Our very idea of "normal" must change. So the term *Ecological Problem* is more meaningful and useful. I capitalize it to emphasize its importance and enduring nature. Throughout this book I will use the term Ecological Problem to denote the array of difficulties that stem from our troubled relationship with the rest of God's creation.

Today, we humans have grown numerous and powerful. Currently there are over seven billion of us on the planet, and we are consuming resources at ever-increasing rates. Our powerful science and technology allow us to exploit and manipulate God's natural world, its creatures, and its processes, to the point that today, for the first time in history, we are actually changing the global ecosphere. "While local overuse of ecosystems has a long history (e.g., overfishing, deforestation, soil erosion), the global human economy has now become so large relative to the regenerative capacity of planet Earth, that it is now for the first time in human history confronting global limits."[21] There is not one corner of the earth's surface that we humans do not live in, visit, or affect in some way—from the ocean depths to the stratosphere, from the Arctic to the Antarctic. As ecologists Peter Vitousek and Harold Mooney put it, "No ecosystem on Earth is free of pervasive human influence."[22] We are changing God's earth and altering his design for it in ways that, because of our limited ecological knowledge, we do not understand.

The reach of our technological power is astonishing. For instance, the Deepwater Horizon oil rig that blew out in the summer of 2010, spilling some eight million barrels of oil into the Gulf of Mexico, was operating in water almost a mile deep and had drilled two miles below the ocean bottom. And oil companies talk of going much farther and deeper in the future to get more oil to support our ever-expanding appetite for energy.

We are impacting ecosystems around the globe through the widespread transport of nonnative species across the world. Global travel

---

20. Tarlock, "Nonequilibrium Paradigm in Ecology," 1140, 1144.
21. Wackernagel and Kitzes, "Ecological Footprint," 1032.
22. Vitousek and Mooney, "Human Domination of Earth's Ecosystems," 494.

and trade have increased exponentially in recent decades.[23] Materials and products are shipped to and fro around the world, and we think nothing of jetting off to faraway places for business or pleasure. But this increased trade and mobility has greatly increased the transportation of animals, plants, and microbes around the planet. All kinds of organisms hitch rides on and in vehicles, people, luggage, and cargo. The legal and illegal pet trade transports numerous exotic species to new places where they can escape and establish themselves. All this has the effect of homogenizing the ecosystems of the world, reducing diversity, and making ecosystems more uniform.[24] It is impossible to know what effects this "eco-globalization" will have on the specialized and unique organisms and ecosystems of God's world.

Global ocean fisheries are under enormous pressure from large-scale industrial fishing. A study done in 2003 estimated that some oceanic wild fish populations have been reduced by as much as 90 percent.[25] A more recent report affirms that overfishing, pollution, habitat destruction, warming, and acidification of the oceans are putting enormous pressure on oceanic ecosystems.[26] Many fisheries are being maximally exploited or overfished.[27] Worldwide, we are "fishing down the food web" as we deplete larger species and turn to smaller ones in order to meet global demand for seafood and feed the rapidly expanding fish-farming industry. This level of exploitation is not sustainable in the long term.[28] The North Atlantic cod fishery collapsed in 1993 due to overfishing.[29] North Pacific wild salmon populations are severely depleted due overfishing, pollution, mismanagement, and damming of rivers used by salmon for spawning.

---

23. The worldwide network of travel and trade requires enormous quantities of energy in the form of fossil fuels—oil and natural gas. In fact, it is the availability of these cheap, portable fossil fuels that has allowed the network's development in the first place. As fossil fuels become more expensive and as pressure mounts to reduce the carbon emissions generated by their use, it is unclear how the worldwide network of travel and trade will adjust.

24. Thompson et al., "Frontiers in Ecology," 22.

25. Myers and Worm, "Rapid Worldwide Depletion," 282.

26. Rogers and Laffoley, *International Earth System Expert Workshop*, 5–7.

27. Garcia and Rosenberg, "Food Security," 2869–71.

28. Pauly et al., "Fishing Down Marine Food Webs," 860.

29. Myers et al., "Why Do Fish Stocks Collapse?" 91.

We are exploiting water resources around the world at or beyond their limits. In his book *Water*, Steven Solomon writes: "Today, man has arrived at the threshold of yet a new age. His technological prowess has reached the point that he possesses the power, literally, to alter nature's resources on a planetary scale, while soaring demand from swelling world population and individual levels of consumption among the newly prospering urgently impel him to use that prowess to extract as much water as he can. The alarming, early result is a worsening depletion of many of Earth's life-sustaining water ecosystems, that, nonetheless, are not keeping pace with the growing global scarcity."[30]

At least forty-five thousand dams have been built worldwide, and more are under construction or planned.[31] In the American Southwest, the once mighty Colorado River is now completely exploited. Some twenty dams have been built on it or its tributaries, and at times during the year, virtually no water arrives at its delta in the Gulf of California. As an exultant President Franklin Roosevelt said at the dedication of the monolithic Hoover Dam in September 1935, we have indeed thoroughly "conquered" the Colorado.

We are also changing the earth's atmosphere. Our modern society is founded on fossil fuels (coal, oil, and natural gas) as our primary source of energy. The widespread burning of these fuels is putting billions of tons of carbon in the form of CO_2 into the atmosphere each year. Although most of this is absorbed back into the earth's terrestrial and oceanic ecosystems, a proportion remains in the atmosphere. As a result, the concentration of CO_2 in the atmosphere has risen from about 315 parts per million (ppm) in 1959 to about 391 ppm in 2011, a 24 percent increase in 52 years.[32] Carbon dioxide levels continue to rise at a rate of about 2 ppm per year, and this rate is accelerating. Global average temperatures have risen about 0.74 degrees Celsius or about one degree Fahrenheit over the last two hundred years or so. Anthropogenic climate change is controversial, and there is much uncertainty, but no matter what your opinion on it is, you have to admit that we

---

30. Solomon, *Water*, 489.

31. Malmqvist and Rundle, "Threats," 138.

32. National Oceanic and Atmospheric Administration, "Mauna Loa Mean Annual Data." According to Vitousek and Mooney, since the advent of industrialization and the widespread use of fossil fuel, atmospheric carbon dioxide has risen from about 280 ppm around 1800 to the current level of 389, an overall increase of roughly 39 percent ("Human Domination," 494).

are changing the atmosphere and that we *don't know* what the outcome will be. Our emitting CO_2 (and other greenhouse gases) into the atmosphere, changing its chemistry, constitutes a gigantic planetary experiment. The unknown results of this could be detrimental to human life and the life of many other creatures. My point here is that we humans are actually doing this; we are literally changing the global firmament that God created (Gen 1:6–8). This is a remarkable example of our power and reach within the created order.

There is evidence that we humans may be using as much as a third of the total worldwide production of plants each year, and that this proportion is increasing.[33] That is to say, we are consuming about one third of the energy contained in all the leaves, stems, fruits, roots, and seeds produced by all the plants on earth each year. The data for this are approximate, but at the very least, they suggest that the rate at which we are using the ecosphere's photosynthetic resources is staggering and may not be sustainable in the long term. We may be beginning to press the limits of the earth's photosynthetic capacity to support us, and we may be inadvertently changing the very relationships and processes that constitute those limits without knowing what we are doing.

These are just a few examples of the ways in which we humans today are impacting the ecology of the planet. Currently the bulk of this impact results from the high living standards and consumer lifestyles of those who live in the so-called developed, industrialized nations of the world, such as those of North America and Europe. But other nations like China, Russia, India, and Brazil are racing to emulate us and are rapidly doing so. As a result, human pressure on God's ecosphere increases steadily. Biologist Peter Vitousek notes that "it is certain that a substantial number of components of global environmental change are now ongoing, and it is equally certain that they are human-caused."[34] Marine biologist Sylvia Earle writes that humans have "engulfed the rest of the living and physical world for food, water, minerals, and materials to build and operate the enormous infrastructure that supports civilization."[35]

---

33. Van Houton and Pimm, "Various Christian Ethics of Species Conservation," 120; Rojstaczer, Sterling, and Moore, "Human Appropriation of Photosynthesis Products," 2550.

34. Vitousek, "Beyond Global Warming," 1870.

35. Earle, *World Is Blue*, 23.

But at the same time, our knowledge of the ecology of the earth has not kept pace. There is much about the earth's ecology and natural systems that we do not understand. We are changing the ecosphere at an accelerating rate, but we don't know what the results will be. I am reminded of a teenage boy I once met at a school a few years ago. Belongings in hand, he was rushing down the hallway at top speed. I stopped him and asked him where he was going. He said, "I don't know where I'm going, but I'm going," and he ran off. We, the human race, are like this boy. We don't know where we're going with God's earth and ecosystems, but we're going.

Scripture teaches that God has given us humans power over his earthly creation (Gen 1:26–28), and he has commissioned us to take care of it (Gen 2:15). We are responsible before the Lord for this. Today, when we are so numerous and so powerful, in order to carry out this responsibility, it is imperative that we have a reasonable understanding of how the system works and how we humans fit into it. We are in control of God's earthly ecosystem, but we are also part of it. Any changes we make will redound to us and our descendants. The Ecological Problem is not just about people, animals, forests, and lands that are far away. It is about you and me, our families and friends, our homes and churches, our companies, jobs, goals, and dreams, in the cities and towns where we live. We are part of the ecosphere; the Ecological Problem is in every way *our* problem.

Christian "Physical Spirituality" and the Ecological Problem

We Christians often assume that reality—our lives and the world—consists of two domains: the spiritual realm and the physical realm.[36] This is usually taken for granted. That is, we automatically assume it to be true without question. We think of these as two different things that are related in certain ways but are separate from one another, or at least separable. To think of reality in terms of these two realms is a worldview called *dualism*. Dualism means dividing something into two parts—like I mentioned earlier about the dualism that sees a human as two parts: body and soul. There are many kinds of dualism around, but here I am talking about *spirit-physical* or *spirit-material dualism*—dualism

36. Probably many secular people think this way too.

that divides reality into two parts, spiritual and physical. We Christians usually think of the physical realm as subordinate to the spiritual realm. It forms a kind of background for the spiritual realm. The spiritual realm might include such things as church, prayer, worship, baptism, communion, evangelism, discipleship, religious experience, and certain personal ethics.[37] The physical realm might include such things as our bodies, food, cars, bank accounts, computers, cell phones, clothes, money, business, houses, shopping, and, last but not least, land, animals, plants, mountains, atmosphere, oceans, and ecosystems. We often think of God as primarily involved with the spiritual realm and not particularly concerned with the physical realm except at certain points like tithing or physical healing. Our spiritual life or "spirituality" concerns the spiritual realm and may (or may not) contact the physical realm at these points. But most of us do not think of our spirituality as involving such things as animals, plants, oceans, standards of living, energy use, consumption, recycling, agriculture, industry, or ecosystems.

I am claiming that we humans are inherently and unavoidably physical creatures. If this is true, then our spirituality must also be physical.[38] There can be no dualism—no two realms. The Bible and the Christian tradition, I believe, recognize this truth. Genuine biblical Christianity does not see the world as divided into the spiritual and physical. It is not dualistic but *holistic*. The spiritual and the physical are two *aspects* of one and the same reality, God's whole creation. And furthermore, as in my second claim (argument 2, above), to be *physical* is to be *ecological*—integrated within the ecosystem. Thus, our *ecological life* is one and the same with our spiritual life. *To live spiritually is to live physically is to live ecologically.* By virtue of our eco-physical nature, our spirituality is inherently physical, dusty, organic, and biological. And if Jesus is Lord,[39] then he is Lord of all—of our eco-physicality and the ecosphere, of plants and animals, of food webs, the water cycle, and the carbon cycle, and of all that I will be talking about in this book. Therefore, for us holistic Christians, *the Ecological Problem is a spiritual problem just as much as it is a physical problem.*

---

37. Today, we sometimes call Christian discipleship "spiritual formation," which, if we think in terms of the two realms, can convey the idea that it does not concern our physical lives or the physical world—except, perhaps, in certain restricted ways. But Christian discipleship or spiritual formation should involve all of life.

38. Van Dyke et al., *Redeeming Creation*, 125.

39. John 13:13; 1 Cor 12:3; Phil 2:11.

Christian Views of Human Nature

Christians hold a variety of views about human nature—what we humans are and how we are put together. This is what theologians call *human constitution* or *anthropology*. Views on this topic have changed over the history of the church and are controversial today among some Christians. Since I am arguing that we should understand ourselves as inherently physical, I want to briefly discuss this here.

As I noted, many Christians see humans as consisting of two parts: a spiritual *soul* and a physical *body*. These two parts are viewed as actual entities that can exist separately under certain conditions—after death, for example. The theological term for this is *body-soul dualism* or *dichotomism*. Some hold that the soul is the essential part of the human being and that the body is of secondary importance. The soul inhabits the body during life on earth, and then at death it departs, leaving the body and the earth behind. The soul ascends to heaven, a nonphysical, spiritual place where God dwells, a place other than earth, usually thought of as being upward, toward the sky. The soul then lives eternally with God in this nonphysical, spiritual state. This is called the doctrine of the *immortality of the soul*. Some Christians hold this state to be a temporary "intermediate state" of the soul and that, at some point in the future, Jesus will physically reappear on earth. There will be a resurrection of bodies, and Christian souls will then return from heaven and reinhabit these new bodies. They will then live for a period of time (a thousand years?) or eternally in a resurrected, physical state with Jesus as ruler of a transformed earth.[40] There are many variations of this view, and the role and importance of the body and the earth vary. But the point is that humans are seen dualistically as consisting of two parts, soul and body, which can be separated from one another, and that the body and the physical world tend to be viewed as of lesser importance.

Another important aspect of dichotomism is that certain human activities or functions may be understood as being done or caused by the soul rather than the body. These vary, but they may include such things as thinking, consciousness, belief in God, hoping, loving, moral decision-making, and spiritual life. These functions are sometimes cited as evidence of our having a soul and as that which distinguishes us from other animals.

---

40. I shall discuss my own view on eschatology, or the last things, in chapter 8.

Introduction: Humans as Dusty Earthlings

Another view sees humans as consisting of three parts: spirit, soul, and body. This is based on 1 Thessalonians 5:23 where the Apostle Paul mentions "spirit, soul, and body." This is called *trichotomism*, recognizing three parts to a human being, but it is probably a less widely held view.[41]

At the other end of the spectrum, some Christians hold that humans are completely physical beings, and there is no immaterial "spiritual" soul at all. Like trichotomism, this is probably a less commonly held view. It is sometimes called *monism*,[42] meaning "one thing," or *physicalism*, meaning humans are strictly physical beings.[43] On this view, the human being is constituted as a unitary, physical being and cannot be divided into parts, as in the dichotomist or trichotomist views. These monists or physicalists don't deny that humans are spiritual beings and have spiritual life, nor do they deny our moral nature. They see spirituality and morality as aspects or facets of holistic, embodied human life.

Now you are probably thinking that I am one of this last kind, a monist or physicalist. Well, not quite. I will share my own (current) view momentarily, but whatever view we take, we must recognize that there are problems (in respect to Scripture, to science, or both) associated with all these views of human anthropology. Perhaps, if asked, many Christians would say they are in between body-soul dualism and monistic physicalism. They would say that humans do have separate bodies and souls but that both are closely connected and both are important.[44]

---

41. For a lucid discussion of these various views of human constitution, see Erickson, *Christian Theology*, 538–45.

42. This monism is not the metaphysical cosmic monism of some religious ideas that sees all of existence as one thing. Here, I am referring to a view that only applies to human nature.

43. There is also a spiritual monism that views humans as strictly spiritual beings—that is, the human is an immaterial being who happens to be using a physical body to manifest himself/herself in the world. On this view, the body is merely a tool or vehicle used by the soul or spirit but does not constitute the person herself. As an explicit doctrine this view is uncommon, but it may occur implicitly more often than we realize, even among Christians.

44. In a recent study done among 250 undergraduate students at the University of Edinburgh, Scotland, 65 percent of respondents "agreed that 'each of us has a soul which is separate from the body,'" and 70 percent believed "that some spiritual part of us survives after death" (Zeman, "Does Consciousness Spring from the Brain?" 294).

As I noted above, the findings of science are making belief in body-soul dualism more difficult to hold.[45] Human actions and experiences that we have traditionally attributed to the soul are being found to involve physical processes in the body and brain, and we are finding that there are many activities that we thought were unique to humans (and therefore to our souls, because we often think that only humans have souls) that we now know to occur in other animals. Be that as it may, the point I want to make here is that, whatever view you hold about human constitution, whether you are a trichotomist, dichotomist, monist, or something else, you and I ought to agree that here and now, in this life, on this earth, we humans *are* physical beings, and as such we *are* integral parts of God's eco-physical creation.

As for my own view, although it continues to change, I cannot avoid the conclusion that we are inherently and fundamentally physical beings. To exist as a human is to exist physically—at least in this age, on this earth, as we are currently constituted. But I also possess an identity, a personality, a mind—or what might be called a plan, pattern, or paradigm—an organizational structure and process that identifies, defines, and expresses me. This identity or paradigm is *not* physical. It is that which persists over time as the physical matter of my body turns over from year to year, and that changes as I go through life, experiencing life, praying, learning, and growing in the Lord.[46] Perhaps it includes the base sequences of my DNA, my memories, certain patterns of neural networks and activity in my brain, ideas and concepts that are embedded in my neural networks and that make me who and what I am and by which others know me.[47] Perhaps this is what the Spirit of God works on to bring me closer to him (Phil 2:13). I would agree to call it a "soul" as long as we understand that it is fully embodied and cannot function in any meaningful way apart from the body. When I die, perhaps it is this identity or paradigm that passes on to inform my resurrected physical body in the new creation (Rom 6:5; Rev 21:1). Thus the

---

45. Theologian Wolfhart Pannenberg notes that during the nineteenth century, "the interpretation of humanity's special place [in nature] as owing to a soul that is united to an animal body became increasingly dubious" (*Anthropology in Theological Perspective*, 27).

46. I have been told that almost all the matter in the human body turns over every seven years. If this be so, then it is this identity or paradigm that ensures that I am the same person today that I was seven years ago.

47. Fuster, *Cortex and Mind*, 6, 53, 251.

Introduction: Humans as Dusty Earthlings

identity of that resurrected person will be me, and not another.[48] How this happens is a mystery that depends wholly on God, and although I do not understand it, I can trust God for it.

I realize that I have not fully explained my anthropology and that, like all anthropologies, it has its problems. (You may be thinking of some right now.) But my purpose is not to argue about human nature or to win you over to my view or to any particular view. What I want you to accept is *our physicality*—yours and mine. Whatever anthropology you hold, I urge that you accept the reality that we are inherently and unavoidably *physical*—and as physical beings we are inescapably *ecological* beings.

The Religious Nature of the Ecological Problem

A few pages back, I made the claim that the Ecological Problem is a spiritual problem. Many secular environmentalists agree. They say that it is not simply a matter of science, technology, politics, and economics—more government laws, more efficient machines, increased productivity, and better management of resources. No, they say it's a religious problem, a spiritual problem, a matter of our attitudes, moral values, and worldview—how we perceive ourselves and the world around us.[49] If we define religion as that set of beliefs that help determine our attitudes, goals, values, and worldview, then I think these people are correct—the Ecological Problem *is* a religious (spiritual) problem.

If this be so, Christians ought to have a lot to say about it. But our contribution has been limited due to several factors, one of which is our belief in spiritual-physical dualism that I mentioned earlier. But if, as I am arguing, our spirituality cannot be separated from our physicality and hence from our ecology, then we Christians agree with these secular folks that the Ecological Problem *is* a religious or spiritual problem. Theologian Ellen Davis notes that in the biblical worldview, "the physical, moral, and spiritual fully interpenetrate one another—in contrast to the modern superstition that these are separable categories."[50]

---

48. See Polkinghorne, *Scientists as Theologians*, 54–55.

49. See, for example, White, "Historical Roots of Our Ecological Crisis," 1206; Oelschlaeger, *Caring for Creation*, 5; Huesemann, "Can Pollution Problems Be Effectively Solved by Environmental Science and Technology?" 285.

50. Davis, *Scripture, Culture, and Agriculture*, 9–10.

In other words, in our bodily eco-physicality, the way we treat God's eco-physical creation and his creatures is a defining feature of our lives as Christians living in God's world.[51] Theologian James McClendon sums it up well: "Our life as Christians is our life as organic constituents of the crust of this planet"[52] (Gen 2:7; Ps 104:27–30).

Some Christians have thought that the Ecological Problem lies in the physical, secular, scientific, or political realm and, therefore, is separate from the "spiritual" or "religious" realm in which Christianity operates. Perhaps, in part, this stems from the separation of church and state so important to our American constitutional system. But if these secular environmentalists are correct that the Ecological Problem is religious, then addressing it is part of the "business" of Christianity. If we believe that Jesus Christ is Creator and Redeemer of all things (John 1:3; Col 1:19–20), then he is Lord of all the earth, its ecosystems and creatures too. As Scripture tells us, God is the creator and owner of the earth—the land and seas and all that is in them[53]—and if we humans are indeed eco-physical beings embedded in the earth's ecosphere, then Christianity has everything to do with the earth, with its creatures, with its ecology, and with the human sciences, technology, economics, and politics that concern these things. This should be a matter of both private and public concern for all Christians. Our personal lifestyles and our public politics should demonstrate our commitment to a holistic conception of life and of God's creation. So for Christians, the Ecological Problem *is* a spiritual-physical-religious problem.

If the Ecological Problem is a religious-spiritual problem, then what is it about our religion or spirituality that is problematic? One extreme view advocated by a small minority of environmentalists is that our very existence is the problem. If we could just rid the earth of humans, they say, the world would be a better place. This, of course, is absurd. Humans are an integral part of God's creation (Gen 1:26, 2:15). As Scripture tells us, and as we shall learn from the science of ecology itself, we are not outsiders in relation to creation; we are part of it, part of the ecosystem. We belong here. Just as we humans would not be human without the earth, so the earth would not be the earth without us humans.[54]

51. Valerio, *"L" Is for Lifestyle*, 37, 46.
52. McClendon, *Ethics*, 89.
53. Gen 1:1; Deut 10:14; Ps 24:1; Rom 11:39.
54. This may sound strange, but it is true. In chapter 6 we will look at the *new ecology* that sees humans as integral parts of the ecosystem.

Another view holds that the Ecological Problem is not religious at all. It is purely a matter of science, technology, and the free market. People of this persuasion are often thoroughgoing materialists or spiritual/material dualists. They may agree that there is a problem and that it lies with humans. But it is not spiritual or religious; it is physical and secular. Humans, they say, have insufficient scientific knowledge of the world, insufficient technological means to control nature (including us humans), and an encumbered market system, weighed down by regulation and government interference. The solution(s) to the problem will emerge spontaneously if we only unshackle the economy, free up industry, and unleash human creativity to develop more efficient engines, better cars, recyclable materials, better ways to exploit resources, genetically engineered organisms (humans?), and so on. In effect, this view is quasi-religious in that science/technology, human intelligence, and the free market system form a kind of tripartite "savior-god" that can solve all our problems and deliver us from all ills. No spiritual or religious change is needed. This optimistic humanism is very common. As a matter of fact, it is the dominant view in American society today, if not the world. It is a kind of unconscious faith that we hold as modern people, and probably, if truth be told, most of us modern Christians believe it too—if not consciously, then certainly in our attitudes and behavior. But this is not consistent with a biblical understanding, and it may even be idolatrous. While human creativity, better science and technology, and better management of markets are certainly needed, this view ignores the deeper spiritual, religious, and moral problems concerning our attitudes, values, priorities, perceptions, choices, worldview, and behavior in relation to God, his planet, and his ecosystems.[55]

---

55. Modern technology has produced birth control pills that allow us to have sex as much as we want without its "normal" biological consequence—babies. I wonder if technology will come up with a pill that will allow us to eat all we want without its "normal" biological consequence—getting fat—or drink all the alcohol we want without becoming addicted or destroying our livers. Or, maybe it will produce some "selfishness" technology that will allow us to be as selfish as we want without any of its "normal" consequences—hurting other people and damaging the earth. Or maybe it will make any number of pills or surgeries or treatments to remove the consequences of our ignorance, foolishness, bad choices, excess, self-indulgence, and so on. I suppose modern science and technology might produce such pills or treatments in the future. But if they do, and we use them on ourselves, I wonder then if we will remain human in any meaningful sense, and I wonder too if it will truly solve the Ecological Problem, or any problem.

Like all major human problems, the Ecological Problem stems from human sin—pride, selfishness, greed, disordered values, idolatry, disregard for God, disregard for other humans and for the creatures of God, and, as I am arguing here, a false understanding about who and what we are and how we fit into God's world.

In what is without question the most influential work ever published in the modern era on Christianity and ecology, the late historian Lynn White, in a 1967 article in the journal *Science*, blamed the Bible and Christianity for the Ecological Problem.[56] His indictment has since been refuted, at least in part, but he was correct in saying that both the cause and the solution to the Ecological Problem are religious in character. "More science and more technology are not going to get us out of the present ecological crisis until we find a new religion or rethink our old one," he wrote.[57] Max Oelschlaeger, a leading environmental philosopher, agrees that the Ecological Problem is, at bottom, religious (or spiritual) in nature: "My conjecture is this: *there are no solutions for the systemic causes of ecocrisis, at least in democratic societies, apart from religious narrative.*"[58] So the Ecological Problem is a spiritual-religious problem with a spiritual-religious solution. And in my view, biblical Christianity offers the best and most plausible solution.

There is a difficulty here, however. If, among the world's religions, Christianity is indeed *the* religion that provides *the* best answer, if it is going to speak prophetically to the world by showing it something of how things ought to be, then we Christians are going to have to demonstrate these things to the world not only in the way we talk but also in the way we live. If it is true that spirituality and physicality are inseparable, then our spiritual commitments are physical commitments and should be visible in our physical (material) lives. Our attitudes, values, and choices expressed in our lifestyles and actions will have to point to the final redemption and healing of the earth that we hope for when Jesus comes again (Matt 19:28; Acts 3:21). If Christianity is a holistic faith, then our redeemed spirituality should be physically evident too. This is not to say that as Christians living more ecologically, we would usher in an ecological utopia or "save the planet." Only God can do such things (Isa 65:17). Our calling is not to "save" the world but to be

---

56. White, "Historical Roots of Our Ecological Crisis," 1203–7.
57. Ibid., 1206.
58. Oelschlaeger, *Caring for Creation*, 5.

faithful to God and his revealed word and to live rightly in his world in a way that is consistent with who we are within his created order. True faith bears tangible (physical) fruit (Luke 6:43–49). As Francis Schaeffer wrote forty years ago: "Surely then, Christians, who have returned through the work of the Lord Jesus Christ to fellowship with God, and have a proper place of reference to the God who is there, should demonstrate a proper use of nature."[59] But Christians have great difficulty with this. Like it or not, for us Christians in our eco-physical life, as in all aspects of life, if we talk the talk, we must walk the walk.[60] Our secular friends recognize the religious-spiritual nature of the Ecological Problem and the religious-spiritual nature of its solution. It is up to us to show them (not just tell them) what this is.

Finding Moral Guidance as Eco-Physical Beings Living in God's Creation

The fourth part of my argument (argument 4, page 4) says that since we are subject to the ecological realities of earthly existence, we ought to be certain kinds of people and behave in certain ways. In other words, on the basis of scientific information from science, I am making moral claims about what kind of people you and I ought to be and how we ought to live.

But modern science, as we ordinarily think of it, can only tell us the *how* of the world—how things work, how things interact, and so on.[61] It can make predictions about what might happen if we do this or that, but it cannot tell us what we *ought to do*. Because science is normally thought of as separate from moral concerns of value and purpose, it can offer no advice as to what moral actions we should take nor what kinds of virtues we should seek to embody. For example, ecologists can tell us that the population of bird species X is in danger of extinction because its population genetics, habitat distribution, and nesting patterns are such that its current population will probably not sustain itself. They can advise us on what measures we might take to improve its chances of survival, such as establishing protected habitats, but the ecologists

---

59. Schaeffer, *Pollution and the Death of Man*, 72.
60. Gushee, "Old-Fashioned Creation Care," 51.
61. Polkinghorne, "Friendship."

cannot tell us whether or not we *ought* to preserve bird species X in the first place, nor *why* we ought (or ought not) to do it. These are questions of value, beliefs, and purposes to which science cannot speak.[62] These must come from somewhere else.

For us Christians, the source of our values, beliefs, and purposes is, or ought to be, Holy Scripture and the theology it embodies. As theologian Loren Wilkinson writes: "Though science has vastly expanded our understanding of the 'is' of the cosmos, nothing in that vast picture of valueless fact gives our explanations or proscriptions any weight—unless there is a context beyond the universe of which we have some knowledge, however imperfect and incomplete. Christian orthodoxy is based on the fact that we do have some knowledge (mediated and imperfect) of such a 'context': it is contained in the texts of Christian Scripture."[63] In other words, *God* has provided to us in Scripture the resources we need for moral life and action in his world. In the example of bird species X threatened with extinction, we would look to the Bible and its theology to answer the question whether or not we should try to preserve it. In Scripture, through stories, poems, speeches, songs, prophecies, letters, instructions, parables, and so on, God reveals to us information about himself, about ourselves, about his world, and about the values, beliefs, and purposes that provide the basis on which we can develop our ethics and determine what kind of people we ought to be and how we ought to live. Moreover, we Christians have an additional source for moral guidance in the person of Jesus Christ, the incarnation of God in the world, the ultimate archetype of humanity as it ought to be.

But it turns out that the Bible also tells us that by studying the natural world (something like the way scientists do) we will find it to be a source of wisdom for living.[64] In fact, by observing nature we

---

62. Conservation biology is a science that by its nature involves the moral imperative that humans ought to preserve other species. This does not, however, defeat the idea that modern science, as we currently understand it, does not make moral claims. Conservation biology's sources for the values that warrant the preservation of species come not from science but from elsewhere—such as personal or religious beliefs. For good discussions of this issue, see Soulé, "What Is Conservation Biology?" and Van Dyke, *Conservation Biology*, 29–55.

63. Wilkinson, "New Story of Creation," 31.

64. See for example Job 38–40; Ps 8:3–4; Prov 30:18–19, 24–28; Matt 6:26, 28–29; Luke 12:24; Acts 17:24.

Introduction: Humans as Dusty Earthlings

can learn something of God. The Apostle Paul wrote, "For since the creation of the world God's invisible qualities—his eternal power and divine nature—have been clearly seen, being understood from what has been made, so that men are without excuse" (Rom 1:20).

The Bible says that God is the Creator of the world and all that is in it. If that is true, then God's authority lies behind the order of creation as we find it. Thus, we Christians can accept information about the order of creation produced by scientific study and use it to help us determine how we ought to live in his world. We can affirm the validity of the sciences (biology, ecology, neuroscience, psychology, sociology, and so on) as sources of information to help us flourish and live rightly within God's creation according to the principles, purposes, and values contained in our theology as it is expressed in Scripture and in the Lord Jesus Christ. In other words, the Christian moral perspective is holistic. We derive our morality not just from Scripture and theology but also from the physical world within the framework of Scripture and theology.

To clarify this point, I will cite two outstanding contemporary Christian thinkers. First, the distinguished American theologian Millard Erickson calls the earth a household, similar to a household or home in which a family might live together. He notes that the word *ecology* is made up of two roots: *eco* and *logy*. *Logy* means "the study of something"—like biology is the study of *bios* or life. *Eco* is derived from the Greek word *oikos*, which means "house" or "household," suggesting the idea that the earthly creation is "one great household" in which all God's creatures live together like a family.[65] The science of ecology, then, is the study of God's *oikos*, his great earthly household, including its living (biotic) members and nonliving (abiotic) components and their relationships. In studying the *oikos*, ecologists (and other scientists as well) can provide us with the "house rules"[66] (what I am calling the patterns, principles, parameters, and limits, or ecological realities) that govern life within the household of creation for all its inhabitants, including us humans since we are members of the household just like all the other creatures. If we believe, as the Bible teaches, that God is the creator and designer of this earth and its ecosphere—or household, to use Erickson's metaphor—then we accept that he has designed into

---

65. Erickson, *Christian Theology*, 511–12.
66. McFague, *Life Abundant*, 72, 122, 208.

it certain principles, patterns, parameters, and limits about how things work in his household and what we ought or ought not do so that all may live together and flourish within it. That is to say, scientific information about how things work in God's earthly household has implications for how we ought to live and behave within it.

The eminent British theologian Oliver O'Donovan proffers the same idea using a different metaphor, the *order of creation*. Speaking against the idea that morality is simply subjective—whatever each of us decides on our own—he says that in creation, there is a "divinely-given *order of things* in which human nature itself is located. Although sinful humans have rebelled against this created order, it still stands and makes its claims upon us . . . The order of things that God has made is *there*. It is objective, and mankind has a place within it. Christian ethics, therefore, has an objective reference because it is concerned with man's life in accordance with this order. *The way the universe is determines how man ought to behave himself in it*."[67] God is the Creator, and he has written into creation an "order of things"—"the way the universe is" as O'Donovan puts it. This "order of things" can be discovered and understood (imperfectly, of course) by us humans through scientific investigation. Christian morality—what we ought and ought not to do—should be concerned with ordering our lives "in accordance with this order." Again, the way the created, ecological world *is* determines how we ought to live and behave within it. Ecological realities have moral implications.

O'Donovan notes that humanity has rebelled against the order of creation, but this does not nullify it; it is still *there*. We Christians, the redeemed people of God, are no longer in rebellion against it (at least, we are not supposed to be). Through Christ our relationship with God has been restored and with it all our other relationships, including our relationship with his creation. Thus, we seek to understand God's created order (ecosphere) through scientific study and to conform our lives, as best we can, to it, seeking within that context to embody the values, norms, and purposes that we derive from Scripture and Jesus Christ. O'Donovan continues, "By virtue of the fact that there is a Creator, there is also a creation that is ordered to its Creator, a world which exists as his creation and in no other way, so that by its very existence it points to God. But then, just because it is ordered vertically in this way,

---

67. O'Donovan, *Resurrection and the Moral Order*, 16–17; italics mine.

it must also have an internal horizontal ordering among its parts . . . It forms over against the Creator, a whole which is 'creation'; and if there is any plurality of creatures within it, they are governed by this shared determinant of their existence, that each to each is as fellow-creature to fellow-creature."[68] Humans are fellow-creatures among fellow-creatures, part of God's creation, and as such are subject to the "shared determinants" of our shared, horizontal creaturely existence, that is, the ecological realities that God has established for his creation, as we, with all the creatures, seek to glorify God.[69]

I want to distinguish my argument here from a more simplistic approach sometimes used by secular environmentalists. Perhaps you have heard them say something like, "Nature is always right." They argue that we should formulate our ecological ethics according to whatever we find in nature. In effect, we should leave nature alone and stay out of it. But what we find in nature is sometimes not what we humans would want to follow morally. For instance, we occasionally encounter destructive storms, fires, earthquakes, and disease. In light of this, I find it hard to understand how these folks can sustain this argument. How can nature be "always right" if it has just wiped out a town in a storm or a species through disease? This is not the argument I am making. I am not suggesting that all that we find in nature is sufficient for building our Christian eco-ethics. Again, the purposes, norms, and values for our ethics come from Jesus, Scripture, and theology. I am drawing on the insights of Erickson and O'Donovan to show that insofar as God is creator and designer of the natural order, this order carries a measure of authority as a framework for our ethics. We apply our Christian ethics within the framework of creation, God's *oikos* (Erickson), or his *created order* (O'Donovan).

I also would like to caution against making the opposite error. Instead of saying nature is always right, we could say that "since nature

---

68. Ibid., 31–32.

69. Wolfhart Pannenberg cites a similar idea in Augustine's concept of sin. For Augustine, the world is ordered hierarchically such that all things come from and return to God. Human sin consisted of an "inversion" of this order by human concupiscence (lust or distorted desire). Our egoism and lust for power lead to our disobedience of "the natural order established by the creator." Augustine evidently held to this same idea of a divinely ordained order of nature that humans should obey (*Anthropology in Theological Perspective*, 94).

is *amoral* it is acceptable for us to be *immoral*."[70] For example, since volcanoes and wildfires are natural and destroy forests and wildlife, we could say it is acceptable for us humans to do the same. Or, since species go extinct due to natural processes, we could say it is okay for us humans to cause extinctions. This is an equally unacceptable approach. Just as we cannot glean morality strictly from nature, so also we cannot glean immorality from it either. Again, our resources for moral norms are Jesus, Scripture, and theology. The context in which we must live out these norms is God's creation.

This brings up the question of natural evil. Nature as we find it embodies great beauty, complexity and grace, and it provides wondrous bounty for the flourishing of all God's creatures, including us. But we also find it sometimes indifferent to the welfare of humans and other creatures—sometimes even brutal and destructive. Storms, fires, earthquakes, and floods can devastate ecosystems and destroy human communities. Competition, predation, parasites, and disease can cause suffering and death. Ecological science has found that events and processes in nature that cause disaster or death can also cause rebirth and renewal. For example, periodic flooding is a normal part of the ecology of most rivers such that some riparian (streamside) plants are adapted to it and cannot reproduce without it. Also, floods deposit silt and mud over large areas, bringing nutrients and renewing the soil. Periodic fires have been found to be integral to the chaparral and coastal scrub ecosystems here in Southern California where I live. The seeds of several plants, for example, must go through a fire in order to germinate. But fires are destructive. Volcanic activity can produce beautiful mountains and renew the land with fertile ash but in the process can cause great destruction, death, and suffering. So we find that floods, fires, and volcanoes are both beneficial and destructive. Even death appears to be integral to ecosystem function. Dead plants and animals are broken down by decay and their nutrients recycled through the system, providing sustenance for new life. Death gives birth to life, and we humans, like all creatures, depend on this reality.[71]

---

70. Hamlin and Lodge, "Beyond Lynn White," 9–10. See also Pickett et al., "New Paradigm in Ecology," 82.

71. Jesus perhaps alludes to this in his comment about the wheat seed dying to give new life (John 12:24).

Introduction: Humans as Dusty Earthlings

Traditionally, Christian theology has viewed natural evil as resulting from the fall of humans (Adam and Eve) in the garden of Eden. Their sin tainted all of created existence and led to the perturbations and evil that we see in nature today. But more recent scientific findings, as I have noted, make it difficult to maintain this view.[72] So we see that the theology of natural evil—whether or not and how nature is fallen, the way in which it is affected by human sin, and the theological meaning of it all—is complex. It presents tough problems for Christian theologians. As our understanding of ecology improves, and as theologians continue their work, I believe our understanding will become more nuanced and mature. Perhaps the suffering and death of Christ on the cross for the sake of his creation (Col 1:20) and the hope embodied in the resurrection of Christ for the future healing and renewal of creation is a resource upon which we can draw as we address this problem. Unfortunately, time, space, and my own limitations do not allow us to explore this further here. We will have to move on. But I recognize the problem of natural evil and that it is germane to the topic of this book.[73]

Finally, we should recognize that our knowledge of God's ecosphere is incomplete. The ecosphere is vast and complex. Ecology is a relatively young science, and its scientific principles are not as well developed as those of, say, physics or chemistry.[74] As we shall see, ecological information is often nuanced and probabilistic. Nevertheless, ecological principles, patterns, parameters, and limits are worked out well enough that we can use them as guidelines for how we ought to live in God's world. As ecologist David Orr has said, "We know enough right now to make far better decisions than we do about wildlife, ecosystems, and landscapes. That is to say, we do not lack for science or data . . . to make better decisions about our 'management of nature' or any number of other things."[75] Furthermore, recent philosophical analysis of the nature and process of scientific investigation has shown that most scientific knowledge is, in reality, probabilistic and uncertain anyway. And if we are honest, we can see that life itself is like that. As ecologists

---

72. Bauckham, *Bible and Ecology*, 160; Polkinghorne, "Kenotic Creation and Divine Action," 93.

73. For an exploration of this issue, discussions of various theologians and their ideas, and an attempted response, see Southgate, *The Groaning of Creation*.

74. Krebs, *Ecology*, 10.

75. Orr, "Retrospect and Prospect," 1350.

study nature, they are learning more and more all the time. Next year they will understand things a little better than they do now. But for us dusty earthlings, there will *never* be perfect understanding or absolute certainty. God is perfect; we are not. God's knowledge is complete and certain; ours is incomplete and uncertain. Perfection and certainty are not "of this world," as we know it. As the Apostle Paul wrote, "Now we see but a poor reflection as in a mirror" (1 Cor 13:12). Besides, God has called us to obedience here and now with the information we have, and the Ecological Problem will not wait for "scientific proof" or "certain" answers. Eco-ethics is unavoidable. We must act because to not act *is* to act. We cannot get off the planet. Like it or not, agree or not, you and I *are* earthlings, and everything we are and do impacts the ecosystem on which we depend and which God charged us to take care of. We must learn as much as we can and do the best we can here and now in order to be faithful Christians and bring glory to God.

A Final Thought

Philosopher Anna Peterson notes that we modern Christians have struggled with our physical nature and our place in the physical world:

> Christian uneasiness about physical bodies has been closely tied to ambivalence about the created world generally. Body and world are physical and transitory in contrast to the spiritual and eternal nature of the soul and of heaven. Christian orthodoxy, however, insists that a benevolent God created both physical bodies and the cosmos itself, which means that material creation cannot simply constitute a trap for spirit. Christian thinkers' efforts to understand the relationship between soul and body reflect the tradition's larger struggle to make sense of the relationship between the spiritual and the physical, between the things of God and the things of the world. These questions raise the ethical questions: What is the value of "this world"? How does God will humans to act in relation to the material creation? Underlying these questions is a central concern of theological anthropology: what is the place of humans, as both physical and spiritual creatures, in the created world?[76]

---

76. Peterson, "In and Of the World?" 242.

Introduction: Humans as Dusty Earthlings

This is what this book is about. As Christians seeking to follow Jesus and advance the kingdom of God in the world,[77] I invite you to join me so that together we can try to "make sense of the relationship between the spiritual and the physical" and "what is the place of humans, as both physical and spiritual creatures, in the created world."

---

77. Stassen and Gushee, *Kingdom Ethics*, 253.

2

Our Separation from Creation

MY ARGUMENT THAT WE need to change our worldview so that we see ourselves as eco-physical beings who are part of God's creation assumes that we do not now see ourselves as such. This, as I have said, constitutes our separation (or alienation) from the rest of creation in terms of our perceptions and our way of life, and it is a root cause of our troubled relationship with creation. Put simply, *our relationship with God's created world is broken.* The theological source of this break is, of course, the fall of humanity into sin in which all human relationships are corrupted (Gen 3), and its remedy is the reconciliation provided through Christ.[1] In this chapter I shall discuss some historical, cultural, religious, and technological factors that widened this rupture. The causes of this are complex, and I do not intend to be exhaustive. My goal is merely to highlight some of these factors in order to help us see how we came to be where we are today.[2]

Before proceeding, let me say that although I see our alienation from creation as key to our troubled relationship with it, I am not arguing for a "return to nature" in which we abandon modern life and technology and take up a "garden-of-Eden-hunter-gatherer" existence. Much of what modern technology has produced is good, but much of this same technology has, perhaps inevitably, contributed to our increasing distance from the rest of the created world. The goal of this book is to help us relocate ourselves within creation such that we can grasp our proper place within it and live in a way consistent with that. While this would not entail an abandonment of technology, it would involve a reorientation of our attitudes toward it such that we become

1. 2 Cor 5:19; Eph 1:10. See Gunton, *Christ and Creation*, 32–34.
2. Northcott, *Environment*, 41.

more thoughtful, critical, moderate, and selective in its development and use. Awareness of our separation from nature and of factors that have contributed to it will help us achieve this reorientation and move closer to fulfilling our call as God's ethical, caretaking species of his earth (Gen 1:26–28), and utilizing our God-given power over nature in a way more aligned with the way of Jesus and with ecological realities.

A number of observers have remarked on our separation from nature that has widened in the modern period. The great nineteenth-century naturalist John Muir writes, "Most people are *on* the world, not *in* it—have no conscious sympathy or relationship to anything about [it]."[3] Muir, who had a rare and profound connection with nature, perhaps could see more easily than others the growing divide between modern humans and the natural world around them. Biologist Calvin DeWitt, a leading evangelical environmentalist, writes, "We have alienated ourselves from the natural processes [of the world]."[4] Ecologist David Orr points to our preoccupation with our constructed world: "We are rapidly becoming an indoor species with fewer people spending time outdoors and with fewer experiential connections with nonhuman nature. Instead we are busy twittering, texting, ipoding, and blogging each other."[5] Anna M. Clark, writer of a recent popular book on ecological care, recognizes this alienation as she reads her Bible: "As I read ancient passages, I am struck by the realization that as wonderful as modern living is, it has severed our bond with nature."[6] Theologian and preacher Elizabeth Achtemeier observes, "The natural world has become strange to us, divorced from our thoughts and vocabulary . . . It is now our own creations that capture our attention instead. Our animals are anthropomorphized into Mickey Mouse and Snoopy. Tiny transistors are more amazing to us than seeds. Concentrating on our manufactured things, we have lost the natural world."[7] The late evangelical theologian Carl F. H. Henry remarked on our assumption that nature is outside of our lives, "wholly different from and inferior" to us. We feel we transcend nature and are "free to manipulate and exploit it to [our] own ends." As a result we tend to "play the role of God over

---

3. Muir, *Wilderness World*, 313.
4. DeWitt, "Eco-Myths," 28.
5. Orr, "Retrospect and Prospect," 1351.
6. Clark, *Green*, 89.
7. Achtemeier, *Nature, God, and Pulpit*, 2.

nature."[8] English theologian Alister McGrath notes that even the intellectual world has divorced itself from nature: "We have lost touch with the world of nature and have constructed our own world in its place. The intellectual pretensions of the modern period have led many to decouple themselves from the natural order."[9]

Ancient peoples by necessity had a closer connection with nature than we do today.[10] They had little control over natural processes and could not protect themselves from the dangers and threats that nature presented to them. Defenseless against diseases, droughts, storms, wild animals, and such, they understood firsthand nature's power. Thus, they conformed themselves as best they could to the natural order as they understood it. They were not inclined to seek the command-and-control approach that we use today.[11] But in the modern era (and probably earlier), at least in the West, as our science and technological power have increased, we have felt ourselves to be more and more independent of nature's rhythms, processes, and power.[12] As a result, we have felt ourselves to be more and more separate from nature and become less sensitive to the ecological realities that govern life on earth.

The Classical World

The Christian church was born and lived for its first several hundred years in the Greco-Roman world of the Near East and Mediterranean region. Greco-Roman culture was not only the context in which Christianity began, it was also a formative source of ideas for the church as well as for Western culture in general, including ideas about human

8. Henry, *God, Revelation, and Authority*, 94.
9. McGrath, *Re-Enchantment*, 186.
10. Hughes, *Environmental History*, 19.

11. Environmental writers have sometimes idealized ancient and aboriginal peoples as living in harmony with nature as opposed to us modern Westerners who both abuse and destroy. But recent research and more thoughtful assessments have largely debunked this myth. In reality, ancient and aboriginal people are humans just like all of us. They exploited nature to the limits of their technology, including some of what we would call abuse. There is evidence they may have deforested some areas, degraded soil, and eliminated species. If they were more eco-friendly, it was probably not because they were somehow morally or spiritually superior but because they had to be in order to survive. See Moran, *People and Nature*, 60–61.

12. Engel and Montagnes, "Environment," 290.

nature, the natural world, and our relationship to it. In early centuries, Christianity encountered a variety of ideas that it had to articulate in terms of its theology and practice. One of these that we have already mentioned was the body-soul dualism of the Greek philosopher Plato and his successors.

Plato

Although conceptions of human nature were numerous and diverse in the Greco-Roman world, a prominent one was body-soul dualism. This idea is very old and probably arose in other parts of the world as well, but it was formally written down and taught by the great Greek philosopher Plato (424–348 BC). Plato's body-soul dualism has had a major influence on Western culture and on Christianity. According to Plato, the "real person" was the soul, the body being little more than a prison in which the soul was trapped.[13] Some of the religions of the day, such as those under the rubric of Gnosticism, adopted Plato's anthropology almost in toto.[14] Thus for them the goal of life was for the soul to escape the prison of the body and the corrupt physical world and ascend to a nonmaterial heaven above where all was perfect spirit. As you can see, this is very similar to the Christian concept of the immortality of the soul that I described in chapter 1. To what degree we should attribute the body-soul dualism of Christianity to Plato's influence and to Greco-Roman ideas is a topic of argument among historians, theologians, and philosophers. But I think there is little question that it had an effect. At any rate, body-soul dualism is still very much alive today in Christian understandings of human nature and perhaps among many non-Christians as well.[15]

Aristotle

Plato was a teacher, and his greatest student was Aristotle (384–322 BC). He rejected Plato's body-soul dualism and redefined the soul as

13. Martin and Barresi, *Rise and Fall*, 15.

14. The Gnostic religions were diverse, but they all believed that true being was immaterial and that matter was evil. The religious task was to escape materiality through spiritual practices and *gnosis*, or knowledge (hence the name *Gnosticism*).

15. Jeeves and Brown, *Neuroscience, Psychology, and Religion*, 109.

a philosophical concept rather than a spiritual substance or entity. The soul—or *form*, as he called it—was the animating essence that gave specific qualities to all living things. Everything, even plants, had souls of varying degrees of complexity. Features such as size, color, shape, growth, life cycle, nutrition, locomotion, rationality, and so on, were all a result of the form, or soul, in the organism.[16] It was the form or soul of an oak tree that made it what it was—a tree-shaped growing thing with leaves, producing acorns. Similarly, it was the soul of us humans that made us what we are—upright, bipedal, fleshy creatures capable of rational thought and free will. The soul was inseparable from the body and died when the body died.[17] Aristotle did not see matter as evil, and his perspective was much more physical and earthy than Plato's. Aristotle was not well known among Christians until the Middle Ages, so his ideas were less influential in the early church. I shall come back to him momentarily when I discuss Thomas Aquinas.

Another important aspect of Aristotle's thought is that he viewed natural things in terms of an instrumental hierarchy. The lower served the higher. The earth and water served plants; plants served animals, and animals served humans.[18] We will see this hierarchical view of the world manifested in the so-called Great Chain of Being, a prominent part of the medieval worldview.

Augustine

Augustine (AD 354–430), bishop of Hippo in North Africa, is perhaps the greatest of all Christian theologians. He was influenced by Plato's body-soul dualism, but in his doctrine, he sought to hold body and soul together as a unified whole.[19] The exact nature of Augustine's anthropology is debated by scholars, but his overall influence on later Christian thinking was toward body-soul dualism.[20] He saw the inward, spiritual life of the immaterial soul as being more important than the

---

16. Perhaps it is a manifestation of Aristotle's brilliance that his notion of soul or form presaged modern genetics. The Aristotelian soul can be likened to an organism's genotype.

17. Murphy, *Bodies and Souls*, 13.

18. Linzey, "Ecological Theology," 264.

19. Martin and Barresi, *Rise and Fall*, 71.

20. Murphy, *Bodies and Souls*, 14.

life of the body. He wrote, "I desire to have knowledge of God and the soul. Of nothing else? No, of nothing else whatsoever."[21] For Augustine, the physical world was in the background where it resides today in most varieties of Christian spirituality in the West. On the other hand, Augustine saw the wonder and beauty of the body and of creation as a whole and affirmed their value in the eyes of God.[22] Augustine's importance in the development of Christian understandings of human nature and the world is hard to overstate.

The Middle Ages

In what we call the Middle Ages, from about AD 500 to about 1500, people generally relied on the church, the Bible, and ancient writers such as Augustine and Aristotle as authoritative sources of information about themselves and the world around them.

In the medieval view, the world was seen as a divinely planned, orderly, harmonious cosmos. This idea was expressed formally in what is called the Great Chain of Being. It depicted all things in a hierarchical system with God at the top, then angels, humans, animals, plants, and nonliving things (water, soil, and rocks) at the bottom. The Great Chain of Being had the virtue of giving an order to the world in which everything had its place, including humans. This order was understood to be divinely ordained and therefore should be respected.[23] One's behavior was to be guided by one's place within this order—all under the dominion of God.[24] The Great Chain of Being was perceived as a hierarchy of the lower things serving (or being used by) the higher. For example, there was a hierarchy among humans with kings at the top, then nobles, merchants, and peasants. Peasants, for example, served their lords and kings, and their lords and kings served (or were supposed to serve) God. Similarly, the earth and its creatures served humans. This view justified human exploitation (and abuse) of the ecosystem, and, indeed, the Middle Ages was a time of significant ecological degradation.[25]

---

21. Augustine, *Soliloques* 1.2.7, quoted in Santmire, *Travail*, 58.
22. Santmire, *Travail*, 62–64; Glacken, *Traces*, 196.
23. McGrath, *Re-Enchantment*, 55.
24. Toulmin, "Cosmology," 29.
25. Hughes, *Environmental History*, 85–95.

In the Middle Ages, the human being was generally viewed dualistically as a composite of soul and body. From their intermediate position in the Great Chain of Being, humans could follow their physical nature downward toward the lower, physical concerns of the earth; or they could follow their soul upward toward the higher spiritual levels leading toward God.[26] "The goal of human existence was often construed in terms of the Christian's pilgrimage within this lowly earth to the beatific vision or union with God in the heavenly Jerusalem above."[27] This motif of ascent upward out of earthly existence toward God remains prominent today among some conceptions of Christian spirituality. One can see a world-denying aspect to this. Earth is a platform or way station on the way upward toward pure spirituality and union with God.

The problem with the Great Chain of Being was that physical things at the bottom were regarded as of lesser value—sometimes even evil, as in the Platonic view. This encouraged "otherworldly, world-denying" attitudes.[28] Involvement with nature was of lesser value than involvement with the spiritual things of heaven. It also implicitly condoned an exploitative attitude toward nature.

In the Middle Ages, Aristotle's writings became available to Christians. Thomas Aquinas (1225–74), a great medieval Christian theologian, was hugely impressed with Aristotle's philosophy and spent much of his life trying to merge Aristotle's ideas with Christian theology. This had the effect of moving the orthodox Christian view toward a more holistic, organic view of human life within the world, effecting perhaps something of a reconnection of humans with their bodies and with nature—at least in some theological and philosophical circles.

An idea that became prominent in the Middle Ages and that has endured until today is the notion that God purposely left his creation rough and unfinished, and it is humanity's divinely appointed task to finish it. On this view, the dominion mandate of Genesis 1:28, in which God commissions humankind to rule over and subdue the earthly creation, is interpreted as authorizing humans to go out and do pretty much what we want to the natural world. Human modification of nature is

---

26. Thomas S. Kuhn, *The Copernican Revolution: Planetary Astronomy in the Development of Western Thought*, quoted in Murphy, *Bodies and Souls*, 43.

27. Santmire, *Travail*, 121.

28. Ibid., 48.

seen as an extension of God's work.[29] Like the Great Chain of Being, this encourages us to think of ourselves as separate from and above nature, and it provides divine sanction for whatever we decide to do to it. It also promotes the command-and-control attitude toward nature, which has become axiomatic in the modern era. This idea is prominent today in certain Christian circles in a radical, utilitarian form.[30] I will comment here that while there is clearly a biblical call for humans to rule over and manage the earth for their welfare *and* the welfare of other creatures, this does not amount to a carte blanche to do as we please. The countervailing ideas that humans are to care for creation (Gen 2:15), that nonhuman creatures and things have value in God's eyes (Gen 1), that God designed the earth as it is and that we should respect it as God's handiwork, and that there are divinely authored ecological principles and limits that we must obey—all this should be kept in mind.

The Reformation and the Renaissance

The Protestant Reformation began in 1517 and did not concern itself much with questions of human anthropology or with the natural world and our place in it. The reformers were preoccupied with salvation and the relationship between God and humans.[31] Body-soul dualism of one kind or another was generally accepted, and the eco-physical world remained in the background. But the Renaissance occurred during the same time period as the Protestant Reformation and led to a revival of interest in classical, pagan Greco-Roman culture and the flowering of human creativity in science, philosophy, and the arts. Its renewed interest in the human body and in nature, coupled with sharp theological disagreements (and wars) among Protestants and Catholics, a

---

29. Glacken, *Traces*, 181.

30. See Beisner, *Where Garden Meets Wilderness*, 13–14, 125.

31. The great reformer John Calvin wrote at the beginning of his *Institutes*, "Our wisdom, in so far as it ought to be deemed true and solid wisdom, consists almost entirely of two parts: the knowledge of God and of ourselves." (1.1.1, 37). In other words, for Calvin, true wisdom concerned only God and humans. The rest of creation is left out. To be fair, Calvin lived within a context where ecological problems were not immediately evident. Also, in other places, Calvin demonstrates remarkable wisdom in relation to creation. See for example *Commentaries*, 125. But we Christians today face clear evidence of serious ecological problems. We should enlarge Calvin's concept of wisdom to include not only knowledge of God and ourselves but also of creation.

fragmentation and decline in religious authority, the discovery of the New World, contact with new peoples, cultures, religions, and lands, the beginnings of modern global capitalism, and new philosophical and scientific ideas all undermined traditional ideas and diminished the credibility of Christian theology. While God became less important, humans (at least some humans) rose in both stature and power in the minds of Western Europeans. This signaled the beginning of the removal of limits on what humans could do to and with nature. Optimism regarding human potential to control and dominate nature grew steadily from this time onward.[32]

The Reformation and Renaissance mark the beginning of the modern period (in which we are now living) and the beginning of ideas such as human command-and-control over nature and human limitlessness that have contributed substantially to our violation of ecological principles and the emergence of the Ecological Problem. The elevation of the human (humanism) eroded the idea that "humanity and nature belonged together or that their destinies were interlocked."[33] The modern idea (delusion) of *human exceptionalism* emerged—the belief that humans are in a separate category from the natural world and are exempt from natural laws and ecological realities, and that humans are capable of controlling and manipulating those laws and realities such that we can effectively ignore them. In short, this period signaled the acceleration of humanity's movement away from nature and toward the ecological predicament we face today.

The work of the astronomers Nicolaus Copernicus (1473–1543), Galileo Galilei (1564–1642), Tycho Brahe (1546–1601), and Johannes Kepler (1571–1630) showed that the earth was not at the center of the universe and that the stars and planets that had formerly been associated with heavenly beings and powers predictably obeyed mathematical laws. Isaac Newton (1643–1727), a brilliant English physicist and mathematician, built on the ideas of the above men and radically changed the way people thought about the world. Newton saw the physical world as consisting of simple, billiard ball–like atoms that interacted within linear, unchanging space according to fixed mathematical laws. That is, the world was thoroughly atomistic and mechanistic in character. His ideas had enormous explanatory power and were quite successful

---

32. McGrath, *Re-Enchantment*, 78.
33. Ibid., 111–12.

when applied in scientific investigation and practical engineering tasks. As a result, whereas previously people saw the universe as a living matrix of beings and things, animated by spiritual forces and sustained by God, they now began to see it as an aggregation of atoms obeying fixed scientific laws—what is called *philosophical mechanism* or the *mechanistic model* of nature. Although neither Galileo nor Newton may have intended it, their ideas led directly to the concept of an effectively dead clockwork universe.[34] Historian Carolyn Merchant sums up:

> The rise of mechanism laid the foundation for a new synthesis of the cosmos, society, and the human being, construed as ordered systems of mechanical parts subject to governance by law and to predictability through deductive reasoning. A new concept of the self as a rational master of the passions housed in a machinelike body began to replace the concept of the self as an integral part of a close-knit harmony of organic parts related to the cosmos and society. Mechanism rendered nature effectively dead, inert and manipulable from without. As a system of thought, it rapidly gained in plausibility during the second half of the seventeenth century.[35]

Today, this mechanistic view is the dominant view in most of the sciences.[36] We will see, however, that, in the twentieth century, at least in the areas of physics and ecology, this notion has been challenged.

Since it was easier for people to think mechanistically about the world outside of themselves than it was to think of themselves in this way, philosophical mechanism tended to widen the perceptual gap between humans and the natural world. "As the world of 'subjective' qualities that the human mind perceives was made secondary to the 'objective' qualities of nature, human subjects began to feel like aliens, unrelated to the cosmos."[37] Humans were inclined to see themselves as separate from nature, at least in terms of their spiritual aspect or their souls. As such they could view nature (and study it) with "objective detachment."[38] This reached its fullest expression in the thought of René Descartes.

---

34. Merchant, *Death of Nature*, 189.
35. Ibid., 214.
36. Midgley, *Ethical Primate*, 95.
37. Shults, *Reforming Theological Anthropology*, 17.
38. Santmire, "Historical Dimensions," 70–71.

René Descartes

The French philosopher and mathematician René Descartes (1596–1650) is an immensely important figure in the development of our modern worldview, and his mind-body and human-nature dualisms have contributed substantially to our modern separation from nature. Fully embracing philosophical mechanism, he saw all living things, including animals and even the human body, as machines. Descartes understood the explanatory power of the mechanistic view, but he wanted to save humans from being reduced to mere machines, so he posited a radical dualism between spirit, or mind, and matter. He conceived of humans as, in essence, immaterial souls—or *minds*, as he called them—inhabiting and using material, mechanical bodies. Descartes was reviving the old Platonic idea of body-soul dualism but in a more radical form. Humans, that is human minds, were immaterial (spiritual) substances that had an existence separate from the physical world, thus promoting a radical dualism between humans and nature.[39] In this position the rational human mind could observe "a machine-like universe from a position of lofty detachment."[40] Descartes saw nonhuman animals as simply machines that functioned according to natural laws; they had no feelings, no personality, no consciousness, and no subjectivity. They were *nothing but* "things" to be studied, used, manipulated, and controlled by human minds for human benefit.[41] Philosopher Stephen Toulmin states, "The crucial change came with the new, mechanical theory of matter and motion . . . As Descartes recognized, if everything in the material world was to be accounted for in mechanistic terms, the capacity of humans to think and act as reasoning beings could be preserved only at a price: that of separating the rational operations of the human mind from the mechanical process of inert matter, and so setting humanity over against nature in a quite new way."[42] Descartes' mind-body and human-nature dualisms have profoundly influenced our modern worldview. In Descartes, the human subject "lost its physicality," and

---

39. Descartes, *Discourse on Method*, 19, 65.

40. Wilkinson, "New Age, New Consciousness," 11.

41. The application of the mechanistic model and Cartesian dualism in modern factory farming and industrial agriculture has been hugely successful in increasing production, but it also results in domestic animals and plants being treated as mere machines.

42. Toulmin, "Religion," 73.

"the physical world lost its spirituality."[43] In subsequent chapters we will see just how wrong Descartes and our modern worldview are, not only about animals but also about us humans. Animals do have consciousness and subjectivity, and humans do have physicality. Humans share many features with other animals. We are not separate beings but are an integral part of the natural eco-physical order.

Francis Bacon

A contemporary of Descartes, the English philosopher and statesman Francis Bacon (1561–1626) is perhaps the principal expositor of what is called the *instrumental view of nature*, the idea that nature exists purely for humans to use for their own needs and desires.[44] Looking to biblical sources, he saw the "dominion mandate" (Gen 1:28) as God's sanction for humanity's absolute control and disposal of nature for our benefit. As Bacon saw it, when Adam and Eve sinned, this control was lost (or at least compromised), and it was the job of science and technology to regain that control so that humanity could utilize nature to its fullest for human well-being. Bacon's instrumental, command-and-control idea, which is sometimes called the *conquest of nature*, was conceived as a means of living out God's redemptive plan for humanity.[45] Thus, Bacon saw nature's value in purely instrumental terms. A secularized version of this view has dominated Western culture for at least the last two centuries and is widely influential today. As we shall see, however, in ecological circles it is being challenged.

At this point, I want to say that an instrumental view of nature is not wholly wrong. It is necessary and correct when kept in proper perspective and context. All organisms exploit their habitats and use other biotic and abiotic resources for their own benefit. Birds use twigs to make nests; gazelles exploit grass, and cheetahs exploit gazelles. We humans use things in our environment for our needs. This is well and good, so Bacon was correct as far as he went. But there are two problems. First, what philosopher Charles Taylor calls "disengaged instrumentalism"[46] sees the self as disconnected from the natural world.

---

43. Martin and Barresi, *Rise and Fall*, 126.
44. Taylor, *Sources*, 232.
45. Clifford, "Ecological Lament," 55.
46. Taylor, *Sources*, 498.

In reality, we are not "disengaged" from the natural world; we are part of it. Radical command-and-control instrumentalism ignores the fact that we are embedded within nature. As history has repeatedly shown us, radical command-and-control can lead to abuses that come around to harm us because of the fact that we *are* part of nature. Even Bacon himself foresaw this problem when he wrote, "nature is only to be commanded by obeying her."[47]

Second, a strictly instrumental, command-and-control view of God's creation and his creatures is not consistent with the biblical view that sees nature's primary value in relation to God. The biblical view sees all things, the whole of creation, first and foremost, as sustained by God and as bringing glory to him. Thus the primary value of all creatures is not their instrumental value for humans (or any other organism), but their goodness and value in the sight of God (Gen 1:31). Nature's usefulness to humans is certainly *a* value of nature, as it is for all creatures, but it is a secondary value.

Nature and the American Puritans

The Puritans were refugees of religious persecution in England and Holland. When they arrived in North America, beginning in the early seventeenth century, they viewed the wilderness that they encountered there as a wild and untamed realm that God had called them to subdue and control. "The Puritans strongly emphasized the text from Genesis which depicts God commanding man to subdue the earth and to have dominion over all creatures [Gen. 1:28]. The Puritans also believed that they were called upon to show the fruits of their election by their works, and so to glorify God."[48] In other words, they carried on the idea that arose in the Middle Ages that they were to "finish" God's creation by subduing the wilderness and building towns, farms, roads, factories, and so on. "Generations were taught by their Churches that nature is properly man's sphere of lordship, given to him by God, and now at his disposal to use, by the sweat of his brow, in order to bring honor to the name of God."[49] This dovetailed well with several modern ideas that

---

47. Bacon, *Great Instauration*, 1:129, p. 370.
48. Santmire, "Historical Dimensions," 71.
49. Ibid.

were prevalent at the time: the instrumental view of Bacon, the pioneer spirit of adventure and entrepreneurship, the burgeoning culture of capitalism and the pursuit of wealth, and the economic theory that individuals acting in their own self-interest developed resources from nature and thus contributed to the greater good of society. For both Puritans and for modern capitalists, the value of nature was merely its "utility" for humans.[50] This is not to blame the Puritans for the modern Ecological Problem. It is merely to show how their attitudes and behavior manifested larger cultural, economic, and religious ideas and movements that were, during that time, feeding modern humanity's increasing separation from nature.

The Enlightenment

The Enlightenment refers to the period from the seventeenth to the eighteenth centuries when the trends that had begun during the Reformation and Renaissance accelerated and matured. In the face of the mechanistic model of nature, some Enlightenment thinkers went in the opposite direction from Descartes and became more materialistic in their understanding of both humans and the world. The English philosopher Thomas Hobbes (1588–1679) rejected altogether the existence of an immaterial mind or soul. Perhaps the first modern physicalist, he believed that humans were material beings only. Thinkers like John Locke (1632–1704) and David Hume (1711–76) may not have denied outright the existence of an immaterial soul, but they emphasized the material aspect of humans and began to think of human activities and identity not in terms of soul and body but in terms of a physically constituted, embodied mind learning from and experiencing the material world around it. This is what is called *empiricism*. It holds that we are products of our upbringing, learning, and experience of the world. The ideas of Hobbes, Locke, Hume and others during this period represent the beginnings of the modern, materialistic, "scientific" view of humans as purely physical beings with which Christianity has struggled ever since. When coupled with philosophical mechanism, this led directly to the reductionistic notion that a human is a purely physical creature whose behavior can ultimately be fully explained (predicted

---

50. Ibid., 72.

and controlled) in terms of the workings of its parts. This view would explode into prominence in the nineteenth century with the advent of Darwinism and biological evolution.

Individualism and the Modern Turn to the Self

With the continued decline of Christianity and the secularization that accompanied it during the Enlightenment, Descartes' idea that we humans, or rather we human minds, are separate from, above and outside of the physical world became secularized and more entrenched. The material world became an impersonal "other' realm to be approached not by building rapport through sympathetic insight, but by manipulation through experiment."[51] God and religion were assigned more and more to the spiritual, nonphysical realm of life, and we humans—or in Cartesian terms, we human minds—became the source of authority (in place of God) for belief about ourselves, the physical world around us, and how we should live in it. As the authority of the church, the Bible, and the ancient sources (Aristotle, Augustine, Aquinas, et al.) declined, people had nowhere to turn except to themselves. The individual person or self or mind, and his or her reason, became the arbiter of truth. This is what historians call the *modern turn to the self*. People turned away from God, away from Scripture, away from the church, away from tradition, and away from ancient authorities, inward toward themselves as the focus of existence and as the source of meaning and significance. This was a profound change in the worldview of Western people of which we today are the heirs. Certainly in American culture, whether we are Christian or not, this understanding of the individual self as the primary entity of meaning and existence pervades our worldview. This individualism has also contributed to our alienation from nature and to the perception that we are exempt from ecological realities. Or, as theologian Larry Rasmussen notes, "While the language of ecology is taken up everywhere, and the metaphors for viewing self, society, and (the rest of) nature are increasingly 'ecological' and 'holistic,' nature outside us is nonetheless absorbed, for all practical purposes, into modern subjectivism."[52] In other words, the human individual (or a particular

---

51. Whitney, "Christianity," 36.
52. Rasmussen, "Ecology and Morality," 262.

social group of like-minded individuals) becomes, in effect, a kind of god—the final authority on truth, meaning, and even reality itself. We become, in effect, creators of our own worlds. As philosopher Anna Peterson puts it, "Human signification creates the world, and any world outside human signification is either meaningless or nonexistent."[53] Secular historian Roderick Nash's influential book *Wilderness and the American Mind* is an example of this modern self-referential approach (and also of philosophical mechanism). He sees wilderness (wild places with minimal human presence) not as complex, interactive systems containing living animals and plants, rocks, soil, lakes, and streams, but as a "state of mind," a mental construct of human minds, a commodity to be bought and sold for human enjoyment and profit.[54] He does not deny the separate existence of nature outside ourselves but insists that its meaning and value are purely a matter of our own human choice and attribution.

The Nineteenth and Twentieth Centuries

In the nineteenth and twentieth centuries, with the flowering of the physical, biological, and social sciences, body-soul dualism was increasingly challenged, as was human-nature dualism.[55] Biological evolution—which said that through purely material evolutionary processes different species gave rise to new species of animals and plants—emerged in the eighteenth century in the ideas of the French thinkers Comte de Buffon and Jean-Baptiste Lamarcke and the British geologist Charles Lyell. In 1859, with the publication of Charles Darwin's *Origin of the Species*, biological evolution took the scientific world by storm. By 1900 evolution was widely accepted as settled "fact" among most biologists. In the twentieth century evolutionary theory has been joined with genetics and population biology to produce what has been called the neo-Darwinian synthesis. Evolutionary thought has challenged Christian conceptions of humanity, origins, and divine action in the world. Darwin saw the world materialistically as a life-and-death struggle for survival under the impersonal "judgment" of natural selection—not a

---

53. Peterson, *Being Human*, 58.
54. Nash, *Wilderness*, 340, 346.
55. Martin and Barresi, *Rise and Fall*, 293.

harmonious, beneficent order designed by a loving God.[56] It supported a purely materialistic view of humans since we too were evolved from other animals as a result of materialistic processes. The responses of Christians to evolutionary ideas have been diverse, fitful, and contentious. Most conservative Christians have tended to reject evolution, while liberals have been inclined to accept it, and there have been any number of responses in between. Seeing evolution as a threat to human dignity and uniqueness, conservative Christians have tended to emphasize human separateness from the rest of creation. While this approach may or may not successfully defend against the challenge of evolution, it has the effect of emphasizing human-nature dualism and of increasing the perceived distance between humans and the rest of creation.

Scientific Challenges to Humanity's Separateness from Nature

Within the scientific community the idea of humans as detached, external observers of nature, while remaining powerful at the level of theoretical and applied science, began to break down in the twentieth century. In physics, the work of Albert Einstein (1879–1955), Werner Heisenberg (1901–76), Louis de Broglie (1892–1987), and others who developed the *quantum theory* of matter and energy showed that human scientific observations of the natural world could be observer-dependent. For example, an electron might behave like a wave or a particle depending on how humans chose to observe it. Human observers were not detached and external to natural events and processes; they were integral parts of them. The way we behave in observing nature can affect how it behaves.

In addition, philosophers of science such as Thomas Kuhn (1922–96) and Michael Polanyi (1891–1976) showed that, like all human activity, science is an immanently social enterprise. Scientists are embedded in a social matrix; their thought and behavior and the development, form, and application of their ideas are strongly influenced by the social contexts in which they live and work. Moreover, the science of *ecology* developed during the twentieth century and showed that ecologically humans are not separate from nature but are unavoidably part of it. As a result of these developments, at least at the philosophical level, the

---

56. Glacken, *Traces*, 183.

Our Separation from Creation

twentieth century has seen a major challenge to the notion of human separation from nature.

Briefly, I should note that the modern era was not characterized exclusively by a mechanical view of nature and human-nature dualism. There have been reactions against this view. The Romantic movement of the eighteenth and nineteenth centuries protested the cold rationality of the Enlightenment and tried to reconnect with nature. People like the English poets William Wordsworth (1770–1850) and Percy Bysshe Shelley (1792–1822), and the Americans Walt Whitman (1819–92), Henry David Thoreau (1817–62), and Ralph Waldo Emerson (1803–82), saw nature as organic and filled with spiritual and moral meaning. They sought to return humans back to their place within the natural order.[57] Today, in our own time, the so-called postmodern movement includes a reaction against human-nature dualism and the mechanistic model of nature. There has been a revival of nature religions and a renewed interest in nature spirituality. Despite these cultural movements, however, the modern view of humans as separate from nature and nature as mechanism still dominates our culture and worldview.

Other Modern Developments

Technology and Industry

Since the eighteenth century, there has been tremendous growth in human technology and industry, providing to us who live in the so-called developed world a degree of safety, comfort, convenience, wealth, and abundance that was unimaginable by our predecessors (including the biblical authors).[58] In our day-to-day lives, most of us live in a world consisting almost entirely of human artifacts. As theologian Joseph Sittler observes, we live in a "made world," and we feel at home in it.[59]

---

57. McGrath, *Re-Enchantment*, 132–34.

58. Elspeth Whitney makes the point that this technical expansion began well before the modern period. "The European West was so successful at technological and economic development that by the fourteenth century Europe had already suffered a series of environmental problems including polluted cities, decreasing grain harvests, and severe pockets of deforestation" (Whitney, "Christianity," 28).

59. Sittler, *Evocations of Grace*, 171.

Food comes from the shelf at the supermarket; water comes from the tap with a turn of the wrist; gasoline comes from the pump at the station; clothes come from the store at the mall; the soil we walk on is covered with wood, plastic, carpet, concrete, or asphalt; in a few hours we fly thousands of miles over mountains, oceans, forests, and storms, in comfortable jetliners at thirty-five thousand feet; and our waste is carried "away" by trucks and sewage systems. Air-conditioning and "climate-control" in our vehicles, homes, and workplaces shield us from weather and seasonal cycles. We surround ourselves with labor-saving machines. The Internet and computers have generated virtual realities almost completely separate from the "real world." With video games we can "live" in virtual worlds and have virtual "out of body" experiences. We have "reduced dramatically—in many places in actuality and everywhere in principle—the contingency of human life in relation to its natural environment."[60] All this has insulated us from the "real" biological-ecological world in which we are, in fact, embedded and upon which our lives and our "made world" depend. We have little or no discernment or understanding of the natural ecosystems that support us—that supply the oxygen we breathe, the water we drink, the food we eat, the clothes on our backs, all the materials of our made world—and that absorb all the wastes we produce. The ecological world exists outside and beyond our experienced human world. Technology in the industrial age blocks out nature from our lives and creates the illusion that nature effectively no longer exists—except when we want to go hiking, camping, and such.

Modern technology and industry have in fact been so successful that they tend to engender the feeling of inevitable progress, guaranteed improvement, and the foreordained, ultimate conquest of all natural obstacles and problems—including the Ecological Problem—leading eventually to utopia. Technology takes on a life of its own and carries us forward into a limitless future. We believe that we can and will eventually reconstruct the world (and even ourselves) such that we will be perfectly and permanently independent of all the contingencies, uncertainties, and perturbations of earthly existence. But technology is a two-edged sword, and success can breed foolish arrogance (Prov 16:18). Besides many good things, technology and industry have also

---

60. Gilkey, *Nature, Reality, and the Sacred*, 170.

produced weapons of mass destruction, pollution, and disease, and they have promoted our separation from God's creation.

The Use of Fossil Fuels

Fossil fuels include coal, oil, and natural gas that we extract from the earth by mining and drilling wells. Since the eighteenth century, we have expanded our use of fossil fuels manyfold until today virtually everything we do depends on them. They are the foundation, the lifeblood of our economy. They have allowed us, for the moment, to live a lifestyle that ignores certain ecological realities—such as the carbon cycle. But like technology, their use has side effects, one of which is a tendency to separate us from the ecological world and facilitate our belief that we can ignore ecological realities.

The Rise of Cities

Over the last century or so, there has been a mass migration of people from rural areas into the cities of the world. Today, over half the world's population lives in cities. But again, city life has the effect of separating us from nature. "Living in urban, technological environments, nature for us is marginalized."[61] We can all too easily forget where our air, water, and food come from and where our waste goes. We are deprived of a "sense of participation in nature."[62] Today, at a time when we most need to reconnect with nature, many of us (including myself) live in cities where we are isolated from it.

Our belief that we are separate from nature, and our modern command-and-control thinking, have led us to build some of our cities against the realities of the local ecology. They are unconnected with or in violation of the ecology of the place in which they are built. A good example is New Orleans, Louisiana, a city built at or even below sea level, near a coastline frequented by hurricanes, next to a large river (the Mississippi) known to flood intermittently. This has necessitated the construction of a huge and expensive system of dikes, levees, channels, pumping systems, and so on, in order to protect the city from

---

61. Taylor, *Sources*, 457.
62. Bauckham, "Joining Creation's Praise," 48.

these perturbations. But as we saw in 2005, when Hurricane Katrina struck the city, our array of devices meant to conquer nature was itself conquered—at great cost to humans (mainly the poor) and to the ecosystem.

Similarly, San Diego, California, the city where I live, contains about 1.3 million people, along with another 1.4 million living in the county and a total of about 18 million living in the greater Southern California region. But Southern California has an arid climate and only enough water to support a fraction of the number of people who are here, especially at the high standard of living that we enjoy—a way of life that requires a lot of water. Our assumed entitlement to virtually unlimited water for our lawns, swimming pools, spas, hot tubs, golf courses, fountains, and lavish landscaping requires massive amounts of water that the region cannot provide. But again, in the true command-and-control tradition, we have overcome this ecological limitation by importing vast quantities of water (over 90 percent of our total consumption) from hundreds of miles away, from the Sacramento and San Joaquin Rivers in the Central Valley and from the Colorado River, via one of the largest and most expensive systems of canals, aqueducts, pipes, and pumping stations in the world. Still, we are dependent on rain and snowfall in the Rocky Mountains (feeding the Colorado) and in the Sierras (feeding the Sacramento and San Joaquin Rivers). Currently San Diego, Los Angeles, and all Southern California have more or less fully exploited these water sources. Some recent years with low rain and snowfall have necessitated temporary water restrictions in the region, but rather than changing our mindset, thinking long term, and reducing our water consumption, aligning ourselves more with the ecosystem in which we live, we are constructing desalination plants along our coast to supply us with even more water. All of this manifests our mentality of separation from the natural world. We are locked into a mindset in which we see ourselves as over the natural world, which we perceive as a limitless storehouse of resources for our use, subject to our command-and-control.

I am not opposed to cities. We need them. They are good. I live in one. But we need to transform our cities so that they are more aligned with their ecology. We are dusty earthlings, part of the ecosystem, and our cities are dusty earth-cities, part of the ecosystem, too. We ought to

Loss of Place

seek to make them more eco-friendly, more aligned with the ecological realities that surround them. More on this in chapter 6.

Some have argued that the only way we can reestablish contact with nature is as particular people living in particular places—as members of local ecological and social communities.[63] Anthropologists have noted that people who have lived sustainably in particular places for thousands of years come to know the climate, the land, and the local flora and fauna of their regions, their homes. They understand their dependencies on the soil, plants, animals, and resources. As a result they learn how to live in a way that is aligned with the principles, patterns, parameters, and limits of their home ecosystem. Some modern farmers, ranchers, hunters, and foresters who work closely with nature also develop these connections. Although we humans have been migratory throughout much of our history, overall we have generally felt more "at home" when we settled down and became residents of particular places. Certainly, much of this sense of belonging is social—being with family, friends, and neighbors in a community—but it is also ecological. People who live for generations in a particular place come to know it well and to love and appreciate it. When this happens, we tend to take ownership and have a deeper commitment for the care and attention required in order to maintain our home over the long term. Not only does the place belong to us, but we belong to the place.[64]

But our individualism, our mobility, our urbanization, our faith in progress, and our emphasis on technology have encouraged an estrangement from place—from hearth and home, and from the local community and ecosystem.[65] We are people on the move.[66] Census data indicate that almost half of all Americans move their place of residence

---

63. Berry, *Dream*, 164. This idea is what is called bioregionalism or localism.

64. This deeper ownership and connection to place might be expressed in the so-called NIMBY ("Not in my backyard") reaction of local residents when a change in their community is proposed.

65. Bratton, *Christianity, Wilderness, and Wildlife*, 239.

66. Social status and prestige are sometimes tied to how much and how far we travel. Statements like "I just flew in from Paris" or "I'm flying to Beijing" usually engender more interest and respect than "I am walking to North Park," the part of urban San Diego where I live.

every five years.[67] Few of us today feel a sense of belonging to any particular place. Our identity is constituted elsewhere—power, job, knowledge, accomplishments, skills, beauty, wealth, and even mobility itself. This too has contributed to our separation from nature.

Conclusion: Where Are We Now? Living on, Not in, God's Earth

We modern Christians have not escaped the influence of these historical, cultural, philosophical, religious, technical, industrial, and demographic movements. Since we are immersed in modern culture, we readily absorb its ideas.

Our alienation from God's natural world is deep and pervasive. It is so much a part of us that we are unaware of it. We have built numerous barriers between ourselves and the natural world, and we live our lives in contrived isolation from the rhythms and movements of God's creation and his creatures all around us. As a result, we know little of God's creation and his creatures, and most of us have little interest in learning about them. Many aspects of our lives are in conflict with the workings of God's ecosphere. If the earthly creation were a symphony orchestra,[68] we humans would be the section that is playing out of tune and out of time—in dissonant contrast with the rest of the orchestra. We would be the ones who don't know the music. But in the tradition of the command-and-control approach, we demand that the rest of the orchestra play to our tune. We continue to try to conquer nature, to beat it into submission. But nature has not been easily beaten, as recent experiences—such as Hurricane Katrina, the Mississippi floods, droughts, storms, and wildfires—have shown. Nevertheless, convinced of our superiority, devoted to modern science and technology, assured of inevitable progress, surrounding ourselves with machines, electronics, and consumer goods, we continue to pursue the modern goal of absolute mastery and control of nature.

In this book, I am claiming that we humans are eco-physical beings. But to say that we are eco-physical is not to reduce us to mere stuff. Indeed, neither is nature mere stuff—inert, meaningless matter.

---

67. Genara Armas, "US Residents Stay on the Move, Census Indicates," quoted in Bouma-Prediger and Walsh, *Beyond Homelessness*, 255–56.

68. Bauckham, *Bible and Ecology*, 78.

Our Separation from Creation

We and nature together are God's singular creation, and as such creation is "enchanted"[69] with God's Spirit (Acts 17:27–28; Col 1:17): it is good (Gen 1); it is valuable; it is complex; it is unpredictable; it is alive; and it is sacred because God made it, owns it, and cares about it (Job 38–41; Ps 104). And we humans, as part of creation, are special because God made us part of that creation and cares about us (Ps 8). As Catholic theologian Charles Murphy writes, "in the biblical world-view . . . nature is not dead but alive. All the animals as well as humanity receive the divine 'breath.'"[70] Will Christians ever recover and live out this biblical worldview and spirituality?

Giordano Bruno was a sixteenth-century philosopher and an enthusiastic supporter of Renaissance humanism. Here he anticipates the modern view of the relationship between humans and nature:

> The gods have given man intelligence and hands, and have made him in their image, endowing him with a capacity superior to other animals. This capacity consists not only in the power to work in accordance with nature and the usual course of things, but *beyond that and outside her laws*, to the end that by fashioning, or having the power to fashion, other natures, other courses, other orders by means of his intelligence, with that freedom without which his resemblance to the deity would not exist, he might in the end *make himself god of the earth* . . . By this means, *separating themselves more and more from their animal natures* by their busy and zealous employments, they climbed nearer to the divine being.[71]

If we ignore the foggy theology in this passage, we see here a concise manifesto of the modern view: humans as detached masters of the created world. Bruno was eventually burned at the stake for his heretical ideas, some of which are evident in this quote. But the worldviews and practices of us modern people—including, if truth be told, most of us Christians—are in agreement with Bruno's idea of human separation from and dominance over creation and our belief in our potential for godlike mastery over it. Such is our modern view, and such is our separation from nature.

---

69. McGrath, *Re-Enchantment*, 188.

70. Murphy, *At Home on Earth*, 95.

71. Giordano Bruno, *Spacio de la Bestia Trionfante*, quoted in Farrington, *Francis Bacon*, 27; italics mine.

3

Human Nature according to Science

BLAISE PASCAL, A SEVENTEENTH-CENTURY French philosopher and devout Christian, said, "It is dangerous to show humans too clearly how much they resemble the beasts, without at the same time showing them their greatness. It is also dangerous to allow them too clear a vision of their greatness without their baseness. It is even more dangerous to leave them in ignorance of both."[1] Ecologically, Pascal's comment goes to the heart of the matter. On the one hand, Scripture teaches that we humans are made in the image of God, gifted managers of God's earthly creation. Science too finds that we are special—in many ways unique among the creatures. On the other hand, Scripture and science both teach us that we are, in every sense of the word, animals—flesh and bone, dusty, dependent beings, equal and in many ways the same as other creatures.

In this chapter, I want to look at what science has uncovered about us humans—what modern science says about who and what we are. By the end of the chapter, I hope that I will have served Pascal's wisdom by showing both our greatness and our baseness as God's gifted, dusty earthlings. In chapter 4 I shall examine what Scripture has to say about this question. I think we will find in this chapter and the next that science and Scripture agree at many points about our human nature and identity. Science is showing us more and more (1) how we are indeed physical beings, (2) how relational we are as physical, ecological, and social beings in the world, and (3) how dependent we are upon God's good earth, his creatures, and his ecosystems for our life and flourishing. All these features of human existence are affirmed by Scripture.

---

1. Quoted in Jeeves, "Nature of Persons," 78.

The plan is as follows. First, I summarize some developments in neuroscience. Second, I examine some of the commonalities we share with other animals. Third, I discuss some aspects of human uniqueness identified by science. Finally, I sum up and then conclude that we humans are the *ethical species*, the only being on the earth possessing the necessary gifts and capacities that permit us to be responsible caretakers of God's earth and his creatures.

Developments in Neuroscience

The last several decades have seen remarkable progress in our understanding of the human brain (and body), how it works, and its relationship to consciousness, behavior, and the social environment.[2] Scientists have identified functions and processes in the brain and body that correspond to and support experiences that we have traditionally understood as "moral" or "spiritual" in character and have attributed to the functioning of an immaterial soul. These experiences and behaviors appear to be embedded in and dependent upon the functioning of neural systems within the brain and body.[3] It appears that "religious and spiritual experiences, like all other human experiences, are grounded in neural processes."[4] Research in biology and comparative animal behavior is also showing that "lines that formerly had seemed to divide animals and humans [are becoming] increasingly blurred, making the question of human uniqueness ever more problematic."[5]

In this regard, I recall the famous case of Phineas P. Gage, whose story is recounted by neurologist Antonio Damasio in his book *Descartes' Error*. Gage was a railroad worker in Vermont who in the summer of 1848 suffered a devastating accident in which a dynamite charge detonated prematurely and blew an iron bar through the left front part of his brain, partly destroying it. Amazingly, Gage recovered from the injury and the wounds healed. Afterwards Gage could walk, talk, perform activities of daily life, and use his hands with his former dexterity, but his character had changed. Whereas before he had been reliable, honest, and efficient, after the accident, he became unstable, irreverent,

2. Jeeves and Brown, *Neuroscience, Psychology, and Religion*, vii.
3. Jeeves, "How Free Is Free?" 105.
4. Jeeves and Brown, *Neuroscience, Psychology, and Religion*, 92.
5. Ibid., 68.

profane, selfish, and unable to control his impulses. In the words of those who knew him, "Gage was no longer Gage." His life became a succession of lost jobs, impulsive behavior, poor judgment, poor social relationships, and alcoholism. He finally died in 1861.[6] Gage seemed to have become morally and spiritually disabled as a result of the injury. Damasio, who has treated a number of patients with brain injuries similar to Gage's, describes the change like this: "Some part of the value system remains and can be utilized in abstract terms, but it is unconnected to real-life situations."[7] Although Gage could verbally articulate his Christian faith and the moral values with which he was familiar, he could not live them out in real life.

What this case and others like it show is that our moral and spiritual life is unavoidably embedded in our bodies, specifically our brains.[8] In fact, our moral and spiritual life is the life of our bodies and brains and vice versa. They are one and the same. When we humans talk, listen, think, imagine, create, learn, laugh, cry, get angry, sin, love, worship, experience God, sense the promptings of the Holy Spirit, make choices, or enjoy the beauty of his creation, we do so only as physical beings—as bodies. There is no other possibility for us. When we pray to our Lord, we pray not only with our minds and our mouths but also with our brains, chests, stomachs, hands, and big toes. We are inherently and pervasively physical—dusty earthlings.

Although Christians in general have been slow to respond to the challenges that neuroscience presents to our traditional understandings of human nature, a few Christian scientists and scholars are confronting them. These include philosopher Nancey Murphy, neuropsychologists Warren Brown and Malcolm Jeeves, Bible scholar Joel Green, geneticist R. J. Berry, theologian/physicist John Polkinghorne, and others. Jeeves compares the challenge of neuroscience to Christian beliefs about the soul to the Copernican revolution of the sixteenth century when we discovered that the earth was not the center of the universe.[9]

Neuroscience is also showing that we humans are inherently and necessarily connected to our environments, constantly interacting and

6. Damasio, *Descartes' Error*, 3–10.

7. Ibid., 11.

8. Numerous other cases like this have been reported in which tumors or injuries to the brain resulted in changed moral and spiritual behavior. See, for example, Burns and Swerdlow, "Right Orbitofrontal Tumor."

9. Jeeves, "Changing Portraits," 23.

exchanging information, matter, and energy with it.[10] Both the growth and development of our brains and bodies as well as their structure and function are dependent on ongoing interaction with our social environment—parents, family, friends, church, coworkers, and community.[11] And, of course, our social environment is completely dependent on the ecosystem in which we live. It provides the oxygen, food, light, resources, waste removal, and so on, without which the very existence of ourselves and our social environment would be impossible. Neuroscience confirms that we are profoundly physical, social, *and* ecological beings embedded within the eco-physical world.

Human Commonalities with Other Animals

Long before the modern era, the great medieval saint Francis of Assisi seems to have understood our membership within creation in a remarkable way. He is remembered for his sense of connection with other animals—birds, fish, lambs, bees, and even worms.[12] In his famous prayer, "Canticle of the Creatures," he refers to even inanimate things as his kindred. Brother Sun, Sister Moon, Brothers Wind and Air, Sister Water, and Sister Mother Earth make up his world, and all bring glory to God.[13] Remarkably, Francis' religious understanding of other created beings as our kindred foreshadowed our modern scientific understanding of ourselves, which has shown that we humans share myriads of molecules, metabolic pathways, membrane structures, organelles, genes, organs, systems, and behaviors with other living things—even bacteria, fungi, and plants.[14] Like other animals, we eat, breathe, urinate, defecate, have sex, reproduce, care for our young, and require a specific habitat and certain ecological conditions in order to live.[15] As Saint Francis saw, we are kin to all living things—and even nonliving things. In this section, we will explore some of the features of this kinship with our fellow inhabitants of the planet.

10. Murphy, *Bodies and Souls*, 56.
11. Fuster, *Cortex and Mind*, 35.
12. Sorrell, *St. Francis*, 42–44.
13. Ibid., 101.
14. Our commonalities with other living things form the basis for our use of them in medical research.
15. Peacocke, *Creation*, 179.

Physical Features

We humans share thousands of structural and functional characteristics with other animals. Anatomically, the human brain and spinal cord are similar to those of other vertebrates (animals with backbones), especially other mammals (animals that have hair and nurse their young). The bone structures of our arms and legs are homologous to those of other mammals, including the fins of dolphins and whales. This similarity extends even to birds' wings and the limbs of reptiles. "All life forms have the same genetic structure and significant genetic overlap . . . Humans are not a biologically segregated species, but rather interrelated parts and products of nature."[16]

Physiologically we have organ systems like other animals. We take food into our digestive system, process it, absorb nutrients, and release waste into the environment. Our circulatory system, consisting of our heart, arteries, veins, and capillaries, is almost identical to that of deer, cats, pigs, and dogs. Our blood contains hemoglobin to carry oxygen like the blood of almost all vertebrate species. Like most other animals, we humans are made up of about 60 percent water by weight. As anyone who has experienced extreme thirst knows, along with food, we need water to live. There is an "inextricable affinity between water and our own essential humanity—not merely with human life, but with a dignified human life."[17]

Cognition (Thinking)

We have traditionally thought of ourselves as *the* reasoning or "rational" creature among the "irrational" nonhuman creatures of earth. In fact, our species name reflects that idea: *Homo sapiens*, the thinking human. But this distinction no longer holds. Other species, such as chimpanzees and dolphins, have been shown to be capable of thinking in terms of perception, categorizing different objects, anticipation of the consequences of an action, abstract concepts, and symbolic thought.[18] Primatologists have found that bonobos (also called pigmy chimpanzees) "are capable of 'thinking about thinking,'" what we call metacognition;

---

16. Nash, "Christianity's Ecological Reformation," 373.
17. Solomon, *Water*, 493.
18. Goodall, *Chimpanzees*, 588; MacIntyre, *Dependent Rational Animals*, 27, 41.

they have reported evidence that nonhuman primates are able to use words and also understand in simple ways how the words fit together (syntax).[19] Investigators taught a chimpanzee named Sarah to "think" in symbolic (non-concrete) terms.[20] Predators like wolves and leopards seem to be able to evaluate prey, conceptualize the environment in which they are working, and make decisions about when, where, and how to attack. Albeit in comparatively simple ways, various animals demonstrate many phases of what we call rational thought.

Social Life (Sociality)

We humans are inherently social creatures. We have long known that many other species of animals are also highly social, and we have, on occasion, used this to our advantage—in the domestication of dogs and sheep, for instance. More recently ethologists (scientists who study animal behavior) have identified complex social structures in the lives of many species including whales, elephants, wolves, dolphins, gorillas, and chimpanzees.[21] Human social life is, of course, much more complex than that of other species, but the point is that social life as such is not unique to humans.

Culture

"The concept of culture, normally exclusively applied to human beings, is now also being applied to chimpanzees by primatologists who have demonstrated geographic variation in the behavioral repertoires of these apes. These variations are described as 'cultural' because these apes are able to invent new customs and then pass them on independently within different social groups, where neither ecological nor genetic factors can account for the behavioral differences manifested between groups."[22] The chimps in Gombe, Tanzania, observed for decades by primatologist Jane Goodall, strip twigs of their leaves and use them to fish for termites through small holes in termite mounds. Chimps in Equatorial Guinea

---

19. Berry and Jeeves, "Nature of Human Nature," 32.
20. Passingham, *What Is Special*, 199.
21. MacIntyre, *Dependent Rational Animals*, 22; Goodall, *Chimpanzees*, 565.
22. Van Huyssteen, *Alone in the World?*, 221.

use larger, thicker sticks and break into the mounds, digging for the termites. Similarly, young orangutans in Borneo spend up to seven years with their mothers in order to learn orangutan "culture"—what to eat, how to behave around other orangutans, how to make a rain umbrella out of leaves, and so on.[23] These culture-like phenomena among these apes are extremely simple compared to human cultures, but they do show in a rudimentary way the possession of customs or traditions that are passed between generations.

Moral-Like Behavior

We humans have traditionally set ourselves apart from other species by noting our concerns about fairness, human rights, virtues, and so on.[24] Sometimes when humans behave in immoral ways we say they are acting in a "beastly" or "animal-like" way, implying that animals are immoral and it is wrong for humans to behave like them—dogs, sharks, pigs, for example. But this is a misunderstanding of the ways of nonhuman animals. They behave as they do within the framework of their own lives and contexts. Philosopher Mary Midgley points out that "every existing animal species has its own nature, its own hierarchy of instincts—in a sense, its own virtues."[25] Wolves, for example, have social systems within their packs with structured relationships and expected behavior on the part of pack members. They are affectionate parents and (in contrast to some humans) seldom kill other animals without good reason.[26] Ethologist Franz de Waal has found that, within the framework of their life and context, many animal species exhibit behaviors that are analogous to human morality.[27] The famous primatologist Jane Goodall observed among the chimpanzees of Gombe greetings, reassurance, food-sharing, altruism, and compassion within the framework of chimp life and social relationships.[28] Dolphins also appear to form social structures within which they communicate vocally, learn, show emotion, interpret one another's behavior, pursue goals, and cooperate

23. Van Dyke, *Conservation Biology*, 51.
24. De Waal, *Good Natured*, 1.
25. Midgley, *Beast and Man*, 47.
26. Ibid., 26.
27. De Waal, *Good Natured*, 218.
28. Goodall, *Chimpanzees*, 360–86.

Human Nature according to Science

in hunting and protection.[29] Wolves, chimpanzees, dolphins, and many other social species have certain ways in which individuals are "expected" to behave according to the structure of their social groups and the requirements of their ecological niches. Thus, one could say they have certain "oughts" and "ought nots," or an extremely simplified form of "right and wrong," analogous to human morality.[30]

There are four things to notice here. First, as we learn more about other animals, namely social animals, we are finding that human morality has its analogues in some of their behaviors. Second, we should avoid the reductionist error. Although we can identify behavior in other animals that is *like* human moral behavior, we must not make the mistake of then saying that human morality may be reduced to merely that. Take altruism, for example, where one animal helps or "loves" another—like parent birds feeding their young or chimpanzees sharing food. Since altruistic behavior in nonhuman animals is generally confined to genetic kindred (kin selection) or tit-for-tat arrangements (reciprocal altruism), we should not therefore conclude that all human altruism (love) is *nothing but* kin selection or reciprocal altruism.[31] While it is true that some of us humans may live on that level, this is not what we would consider well-developed, moral human life—especially when we consider the ethics of Jesus and his gospel (for example, Luke 6:27–36).[32] Third, it appears that the moral-like behavior of nonhuman animals always takes place within a given species. Animals behave in moral-like ways only toward their own kind and often only toward their own particular family or genetic kin. By contrast, humans exhibit moral behavior outside their family groups and even toward other species. For example, in 2010, during the Deepwater Horizon oil spill in the Gulf of Mexico, humans rescued oil-soaked birds, cleaned them, nursed them back to health, and released them—all done with no benefit to themselves. This kind of interspecific (between different species) caring behavior is not seen in other animals. Fourth, although some animals may exhibit moral-like behavior, this does not mean that they are indeed

---

29. McIntyre, *Dependent Rational Animals*, 27.

30. Ibid., 21–51.

31. Current theories of altruistic behavior in animals (and humans) in evolutionary biology are quite complex. See, for example, Nowak and Sigmund, "How Populations Cohere."

32. Green, *Body, Soul, and Human Life*, 79. See also Williams, *Doing Without*, 186–87.

moral beings in the way that we humans are or that they can be praised or blamed for what they do (or do not do). "The appearance in different species of self-sacrificing behavior does not disclose for us the basis of those behaviors."[33] So far as we know, only humans are capable of thinking reflectively about their behavior. They can ask the questions: What am I doing? Why am I doing it? What should I be doing? Animals can learn moral-like behaviors or perform them instinctually, but they do not appear to be self-conscious and cannot reflect upon their own behavior and evaluate it in light of what we would call moral norms, principles, or virtues. As C. S. Lewis reminds us, "So far as we know beasts are incapable either of sin or virtue."[34] So although many animal species exhibit behavior that is analogous to human morality, or is moral-like, true moral life is unique to us humans.

Ecological Life

Going all the way back to Aristotle, naturalists and biologists have recognized that ecological factors shape the lives and behaviors of all living things. All plants and animals are embedded in ecological contexts and processes and live within the parameters and constraints of their earthly locations in terms of temperature, pressure, chemical milieu, climate, seasonal variations, terrain, the presence or absence of predators, parasites and prey, availability of light, food, water, and so on. Humans are the same. Today in the modern world, it is difficult for us humans to recognize this fact since we have achieved such far-reaching control over nature and have so thoroughly modified our environments such that, for the time being, we screen these factors out of our lives. But the enduring reality is that we, like all living things, are an ecological species. We too are embedded in and dependent upon our ecological contexts. Our lives and behaviors are defined and circumscribed by the ecology of the places in which we live. Scientists have also demonstrated that among social animals, their social arrangements are molded by ecological realities as well, such as the availability of water and food, seasonal cycles, and so on.[35] Similarly, humans who live in different ecological environments have developed different social structures and

---

33. Ibid., 78.
34. Lewis, *Problem of Pain*, 129.
35. Strier, "Primate Behavioral Ecology," 19.

customs. Here again, like other social animals, our sociality as a species is dependent upon the ecological context in which we live. Theologian Joseph Sittler puts it this way: "I am what I am not only as one with, among, and in self-forming transactions with men; I am who and what I am in relation to the web, structure, process and placenta of nature."[36] Humans can only live as part of an ecosystem.[37]

Human Uniqueness—Differences from Other Animals

In the Bible, we humans are set apart from other creatures in that we are created "in the image of God" (Gen 1:26–27). I shall discuss this in chapter 5, but does science identify any features of human life that set us apart from other creatures? In other words, what is it about humans that justifies my writing this book to you, a human, instead of to, say, dolphins or chimpanzees? Scientifically speaking, why did God give dominion to humans and not to some other animal? What makes us humans the *special species* who can transcend our natural instincts (1 Pet 2:12), limit or even deny our self-interest, care for others—even other species—and be held responsible by God for our character, our behavior, and for the way that we treat ourselves, other people, and his creation? In this section I want to offer some scientific answers to these questions.

Brain Size and Structure

We humans have by far the largest brain for our body size of any animal. Chimpanzees are the animals most like us humans, but measurements show that our brain is 3.5 times larger than it would be if we were simply big chimps.[38] In other words, based on proportional brain size, we are not simply big apes, we are humans. Furthermore, the frontal lobe, the part of the brain involved in higher functions, is about 1/3 larger than expected if we were simply big apes.[39] Even more to the point, the front part of the frontal cortex (the prefrontal cortex), which is the part of the brain most responsible for higher human thought,

---

36. Sittler, *Evocations of Grace*, 196.
37. Conradie, *Ecological Christian Anthropology*, 100.
38. Passingham, *What Is Special*, 33, 193.
39. Ibid., 33.

creativity, imagination, morality, and spirituality,[40] makes up about 29 percent (almost 1/3) of our brain, whereas it makes up only 17 percent of a chimp's brain and much less in lower animals.[41] These differences amount to a huge increase in the volume of our brains, and hence in the number of cells and the complexity of interconnections and networks that are possible. Neuroscientist Richard Passingham remarks that "the dramatic finding is that the gap in relative size between the human and chimpanzee brain is larger than the gap between the chimpanzee and shrew."[42] Also, the microstructure of the human brain is more developed and complex than that of other animals.[43]

Relationality

Humans must exist within a web of theological, social, physical, and ecological relationships—with God, with other humans, with other creatures, with the land, sea, atmosphere, and the ecosystems in which we live. We cannot survive and flourish outside of these relationships. This is so obvious and basic to our existence that we rarely think about it. In order to function as humans, we must have adequate food, housing, space, family, friends, and so on. Numerous scientific studies and historical events confirm that intact social and ecological relationships are essential for human health and happiness. As theologian James Nash writes, "Humans are internally related creatures—dependent from cradle to grave on the cultural and ecological conditions that shape all their perspectives and possibilities. They are parts and products of collectives—not only of family groups and communities but also of ecosystems. Humans are social animals, as mainstream Christian traditions have always understood, but they are also ecological animals, as the traditions rarely noticed."[44] Theologian Joseph Sittler echoes Nash's words: "Man has what being he has among things, and with things . . . He is *an ecological entity in relation.*"[45] All creatures are relational, but

---

40. Fuster, "Frontal Lobe," 378.
41. Ibid., 374–76.
42. Passingham, *What Is Special*, 57.
43. Ibid., 196.
44. Nash, "Christianity's Ecological Reformation," 274.
45. Sittler, *Evocations of Grace*, 79; italics mine.

our human relationality is so highly developed, pervasive, and complex that it renders us unique among all the creatures of God's earth.

Social Life (Sociality)

We humans are inherently social beings. There are many other social animals, such as chimpanzees, wolves, and dolphins, but no other animal exhibits the extremely high degree and complexity of social life that we humans do. As Aristotle noted long ago, we humans are "political animals." Who and what we are—our identity and personality—are massively influenced by our social experience, context, and relationships. As children growing up, we require a suitable, supportive social environment in order to develop normally. As shown in so-called feral children who grow up without adequate social support, even the development of the physical structure and function of our brains depends on this. We humans can cooperate with others and share ideas, intentions, and goals to a degree and level of complexity not seen in other animals.[46] Much of our day-to-day living is focused on our social relationships. Social networking on the Internet has exploded in recent years, and we even have a multibillion dollar gossip industry that exploits our sociality for profit. To be human is to be social; it is in our DNA. Social awareness comes naturally to us; unfortunately, ecological awareness does not. This is why we need to intentionally develop our ecological consciousness (more on this in chapter 8).

Speech, Language, and Communication

We humans are unique in that we possess highly developed, complex, symbolic language, both spoken and written.[47] Language supports the sociality that I just mentioned. We use not only words but also body language, clothing, ornamentation, possessions, written words, song, and any number of tools and devices as means of communication. We possess a vast array of overt and subtle facial expressions and movements that convey meaning.[48] Beyond this, we are capable of "inner speech"—of

---

46. Passingham, *What Is Special*, 30.
47. Berry and Jeeves, "Nature of Human Nature," 10.
48. Jeeves and Brown, *Neuroscience, Psychology, and Religion*, 74.

Dusty Earthlings

carrying on conversations with ourselves in the form of trial scenarios, thinking about our thoughts, or running commentary on our lives.[49]

Other animals such as lions, dogs, and chimps use facial expression but with nowhere near the complexity and subtlety of humans. Birds, chimps, whales, and dolphins use language-like behavior or sounds to communicate, but so far as we know, human communication is vastly more complex, abstract, diverse, and nuanced than that of any other animal. The structure of our larynx, tongue, mouth, head, chest, and throat seem to be unique, adapted to the production of an incredible array of sounds, clicks, tones, grunts, and even beautiful song (that can rival the nightingale).[50] Sophisticated, complex language and communication are unique to our human species.

Cognition and Metacognition

As we noted above, various nonhuman creatures seem to be capable of cognition or thinking, but humans have an extraordinary capacity to think in varied and complex ways. Metacognition basically means *thinking about thinking*, and this we can do with remarkable power. We are capable of complex, creative, abstract, self-transcending thought. With our voluminous brains, we can engage in marvelously elaborate and diverse cognitive and emotional activity. A few other animals have been found to demonstrate self-awareness and self-recognition. They appear to be able to "represent themselves in their own mental processes" and to think about thinking.[51] Although these animals show evidence of metacognition, only humans can do it at highly complex and sophisticated levels. We are capable of what we might call *transcendent self-consciousness*. We can think about ourselves and what we are doing from a perspective outside of ourselves. We can picture ourselves in the world and in relation to other people and things, ask questions about this, examine the past and imagine the future, and formulate theories and predictions about our behavior or the behavior of others and how the world might respond to various actions.[52] Our capacity for

---

49. Passingham, *What Is Special*, 8–9, 30, 80.
50. Ibid., 106.
51. Jeeves and Brown, *Neuroscience, Psychology, and Religion*, 122.
52. Smith, *Moral, Believing Animals*, 120–3.

self-transcendence is, in my view, a key feature of our humanity—our moral and spiritual character and our ability to relate to God.

These capacities give us magnificent powers of self-awareness, death-awareness, complex reasoning and emotion, abstract imagination, creativity, moral life, and spirituality. Theologian Sallie McFague ironically remarks: "As the symbol-making creatures par excellence, we have the peculiar, painful, and wonderful distinction of knowing that we know."[53] But it is important to remember that whereas our thinking capacity is orders of magnitude more complex and powerful than that of any other animal, it remains the activity of eco-physical creatures—us humans embedded within this earthly creation. Our thinking, as wonderful as it is, remains that of "one species of animal."[54] "The simple truth is that humans cannot think if they cannot breathe and if they do not eat nutritious food and drink clean water."[55]

Technology

Compared to other animals, humans have remarkable manual skill. The human hand with its opposable thumb is a unique and amazing instrument. With it we write great literature, create beautiful art, build skyscrapers, show love, and express diverse emotions. With our hands and our brains, we create complex tools and technology. Chimps can strip twigs to fish for termites, crush leaves to use as a sponge, and use stones to break nuts—as do some monkeys and birds—but "their survival does not depend on tool use." We modern humans "depend critically on the sophisticated technology that [we] have developed to allow [us] to adapt to different climates, produce a variety of food and defend [ourselves]."[56] As we have seen in our modern nuclear/electronic age, we can expand our technology with great speed and imagination, but it remains to be seen if we will further differentiate ourselves from other species by appropriately controlling and limiting our technology when ecological realities require it.

---

53. McFague, *New Climate for Theology*, 43.
54. MacIntyre, *Dependent Rational Animals*, 5.
55. Conradie, *Ecological Christian Anthropology*, 117.
56. Passingham, *What Is Special*, 91.

Agriculture

Closely related to technology is our capacity to control and use other organisms to our advantage. We know of a few other organisms that carry out agriculture-like activities. Leafcutter ants, for example, cultivate a fungus in their nests for food. But we humans have domesticated and controlled a vast array of plants and animals that we breed, cultivate, and manipulate for food, clothing, protection, construction, companionship, and many other things. And we continue to expand and develop this activity with our worldwide trade in exotic animals and genetically modified plants and animals.

Complex Culture

Thanks to our bodies, our brains, our social nature, our capacity for language, communication, and abstract thought, our technology, our agriculture, and the diverse ecosystems in which we live, we have developed a wondrous array of cultures with science, literature, arts, political institutions, traditions, philosophies, religions, and much more. We teach our culture to our children, handing it on from one generation to the next. As I noted previously, a few other species have been found to possess features that could be likened to human culture in the form of simple techniques that are passed from generation to generation. As we saw, chimps pass along termite-fishing techniques to their offspring, and orangutans spend years with their mothers learning how to live in the rainforest. Likewise, cheetah cubs learn hunting methods from their mothers, and dolphins pass on "patterns of giving and receiving" from adults to offspring.[57] Many other species learn foraging behaviors, social rules and hierarchies, and other "traditions" through social contact with their elders, and there have been a few instances where a species has improved a practice over time. Japanese macaques were observed to wash potatoes, then began washing other food items, then began digging water holes to catch rainwater to wash their food so they did not have to walk so far to get to a water source.[58] These and other species may show culture-like features and characteristics, but it all pales compared to the diversity, complexity, beauty, and sophistication of human culture.

---

57. MacIntyre, *Dependent Rational Animals*, 26, 82.
58. Høgh-Olesen, "Homo Sapiens—Homo Socious," 243, 244.

Future Planning

Animals can plan for the future. Arctic foxes bury food for winter, squirrels store away acorns, and bears fatten up for hibernation. But humans seem to be unique in that in our minds we can imagine alternative future scenarios, do trial runs, evaluate reasons for different actions, chose among alternatives, cooperate with others, and then alter our plan when conditions change.[59] Furthermore, we can do long-term planning—imagining a future many years or even centuries hence.

Free Choice or "Willed Action"[60]

Behavioral biologists have found that several different species of animals appear to be able to make simple choices. A wolf pack may break off the pursuit of an elk herd, deciding that it isn't worth the effort. A cheetah may choose a particular individual as a target prey among a herd of gazelle, thinking that it might be easier to capture. A female peacock may select a particular male for a mate because she finds him attractive in various ways. But we humans appear to be unique in being able to make sophisticated and complex choices while aware that we are doing it. We are (or at least can be) aware of our intentions, but we are unsure if this occurs in animals. We can also "veto our intentions at will."[61] We can inhibit behavior and voluntarily limit various aspects of our lives if we choose. We can seek secondary rewards and set long-term goals.[62] While other animals may make simple choices, we humans possess a degree of freedom that goes far beyond this in power and complexity.

Morality and Ethics

As we have seen, moral-like behavior has been observed in many non-human species, but, so far as we know, we humans are the only species capable of true moral life—that thinks and communicates about what kind of animals we ought to be and what we ought to do. "Human

---

59. Passingham, *What Is Special*, 56, 155.

60. Ibid., 10. "Willed action" is a term coined by the early twentieth-century psychologist William James.

61. Ibid., 149.

62. Ibid., 151.

persons nearly universally live in social worlds that are thickly webbed with moral assumptions, beliefs, commitments, and obligations."[63] We are the only species with *ethics* where we study, think, and communicate about morality. Far beyond the capacities of any other species, we can evaluate our actions in light of principles and standards of right and wrong that we have set for ourselves. We can anticipate the effects of our actions on other humans, on other creatures, and on the ecosystem. Primatologist Franz de Waals notes, "Even if animals other than ourselves act in ways tantamount to moral behavior, their behavior does not necessarily rest on deliberations of the kind we engage in. It is hard to believe that animals weigh their own interests against the rights of others, that they develop a vision of the greater good of society, or that they feel lifelong guilt about something they should not have done."[64] De Waals comments in another work: "humankind's uniqueness is embodied in a suite of features that include ethical behavior and religious belief."[65] From the perspective of science, it appears that no other animal seems to struggle with good and evil, right and wrong, or reward and punishment as we do. In both a qualitative and a quantitative sense, morality is a unique property belonging only to humans.

Religion

We humans appear to be the only species that asks questions about ultimate existence, origins, meaning, values, and death. Of course, we have not, as yet, been able to probe deeply into the "thoughts" or "conversations" of, say, dolphins, wolves, or chimpanzees, but it seems quite unlikely indeed that they ever ask (or discuss) such questions. We seem to be the only species that possesses systems of belief that offer answers to these questions. In other words, *we are the religious species.*[66] This is dependent on our capacity for metacognition, and, of course, we carry out such reflections only as embodied beings. Humans contemplate death and nonexistence and wonder about origins, life after death, look

---

63. Smith, *Moral, Believing Animals*, 8.
64. De Waal, *Bonobo*, 216.
65. De Waal, *Good Natured*, 218.
66. Perhaps the first person to cite religion as a distinguishing mark of the human species is Pico della Mirandola, a fifteenth-century thinker in Florence, Italy (Martin and Barresi, *Rise and Fall*, 113).

back to life before birth, and even wonder about what might have been before the whole universe existed or what might be after it ends.

Migration

Jared Diamond, in his book *Guns, Germs, and Steel*, calls humans the "migratory species." According to current theories, humans are believed to have originated in Africa, but we did not remain there. Fairly early in our history, we began moving, crossing land and sea, until today there is no place on earth that humans do not inhabit or visit.[67] Each year the Arctic tern migrates over twenty thousand miles, from Greenland to Antarctica and back. Ornithologists (bird specialists) think that some individual terns may cover a trip to the moon (238,000 miles) during their lifetimes. But we humans put Arctic terns to shame. With our cars, trains, and planes (all powered by fossil fuel, of course), we travel billions of miles each year, and some individual humans may travel millions of miles in their lifetimes. As we noted in chapter 2, many of us are continually moving our place of residence. Again, although other species such as the Arctic tern migrate remarkable distances, we humans move and migrate on a level that is orders of magnitude greater than any other living thing on the planet. Diamond is right: in our restless movement and travel, we are truly *the migratory species*.

Fire

We humans appear to be the only animal who can use (and abuse) fire for our own purposes.[68] We use fire for a multitude of things: keeping warm, cooking our food, running our cars, trains, and planes, burning forests to provide pasture or farmland, smelting iron or copper to make tools, or burning the homes of our enemies. Fire has been an enormously powerful tool in our hands for the control and modification of the world around us, including the ecosystem. We have become radically dependent upon it. In fact, we could say that fire in the form of the burning of fossil fuel "is at the heart of today's world."[69]

---

67. Diamond, *Guns*, 37.
68. Simmons, *Global Environmental History*, 35.
69. Ibid.

Love

I list this last because it is, perhaps, that feature of our human existence that is most like God. By love I mean the capacity to carry out an action on behalf of another being without any benefit to ourselves. Evolutionary biologists have struggled with this aspect of human behavior. They call it altruism, and, as I noted previously, they have come up with two theories to explain it—kin selection and reciprocal altruism. In kin selection one animal may help another if they share genes; thus the animal is, in effect, helping to preserve and pass on its own genes to the next generation, which, according to evolutionary theory, is the overriding goal of life. For example, in wolf packs, related members—cousins, aunts, and so on—help rear young who possess some of their genes. In reciprocal altruism, animals do tit-for-tat favors for one another. For example, among vampire bats, on a given night one bat may share its blood meal by regurgitating part of it for another bat who failed to get a blood meal that evening. But the recipient is expected to pay it back in the future. The bats remember who owes whom and punish those who do not pay up. We humans engage in kin selection and tit-for-tat altruism, too. We care for our families and children, and we do favors for our friends, expecting favors in return. But beyond this, and with surprising frequency, we help one another when there is no apparent advantage for our families (genes) and no expectation of being paid back. In fact, as I have noted, we even help nonhuman species when there is no clear benefit for us. This kind of interspecific (between different species) altruism has been observed in nature only when both species obtained some benefit from the exchange. For example, a species of acacia tree in South and Central America houses a species of stinging ant in its hollow thorns. The tree provides the ants with shelter and food in the form of sugar and protein secretions, while the ants attack anything that threatens the tree, trim vegetation that grows close to it, and eat any fungi that grow on it. This is called *mutualism*, and there are numerous examples in nature. But humans seem to go beyond this by helping other animals apparently only for the sake of those animals. Recall the rescue of oil-soaked birds by humans during the Deepwater Horizon oil spill that I mentioned earlier. Conservation biologists devote their lives to preserving other species. Perhaps in this way they exhibit the most remarkable expressions of love there are. As theologian Christopher Southgate observes, "The human animal has access to an extent of freedom and self-transcendence that goes vastly beyond what is present in other animals."[70] Perhaps self-giving love, even toward other species, in the image of God may be the ultimate in "human" uniqueness.[71] In fact, some have suggested that the behaviors promoted by Jesus go against our evolved propensities and "defy natural selection as much as [they do] cultural conventions."[72]

Table 3.1
Humans and Other Animals: Commonalities and Differences

Commonalities	**Differences**
Biochemistry	Brain size and structure
Anatomy	Relationality
Physiology	Social life (sociality)
Cognition (thinking)	Speech and language
Social life	Cognition and metacognition
Culture	Technology
Moral-like behavior	Agriculture
Ecological life	Complex culture
	Future planning
	Free choice (willed action)
	Morality and ethics
	Religion
	Migration
	Fire
	Love

Humans—The Ethical Species

The features we have in common with other living things and the differences that constitute our uniqueness are summarized in Table 3.1 above. As you can see, I have listed some things as both differences and commonalities. For these, the distinction is a matter of sophistication and degree. We are unique animals who are capable of complex communication, language, rational thought, self-transcendence, moral

70. Southgate, *Groaning of Creation*, 71, 84.
71. Ibid., 72.
72. Green, *Body, Soul, and Human Life*, 79.

choice, creativity, and love. This is the basis for all the communicating we do in which we seek to persuade one another to be certain kinds of people, or to do or not do this or that. It is why God gave his Bible to us and not to chimpanzees. It is why I am writing this book to you as my fellow human, seeking to persuade you about ideas, perceptions, and behavior. I am engaging with you in ethics, which is, as we have seen, a uniquely human activity. As sociologist Christian Smith puts it, we are moral animals.[73] We are uniquely the *ethical species*. More than anything else, this is what truly sets us apart from all other species on earth.

But, at the same time, we are animals—biological beings. We share all aspects of biological existence with all other earthly life. While we possess many characteristics that, taken together, establish our uniqueness, these do not remove us from the categories of "animal" or "organism." All the unique human traits that I have discussed do not lift us out of the ecosystem. They affirm our place within it. We remain one species among many. In a real sense, we are *primus inter pares*—first among equals. Yes, we are unique, but we are unique *animals*.[74] "Human uniqueness can be adequately addressed only if we take our own animality and embodied personhood seriously in both theology and the sciences."[75]

Bible scholar Richard Bauckham offers the paradoxical suggestion that our uniqueness is embodied in our commonalities with other creatures: "The distinctively human emerges from a deep continuity with the rest of nature and is by no means easy to define . . . The recognition that the distinction between humanity and the rest of nature is not an absolute one has become very obvious through modern science, but has often been part of ordinary human reflection on humanity's place in the world, and is in fact present in the Genesis 1 account of creation."[76] Similarly, theologian Ernst Conradie observes, "We can only identify and describe ourselves as human beings by making a distinction between human and nonhuman species. Our identity as humans is therefore constituted by nonhuman animals. We are human only because others are nonhuman."[77] This is an important insight. Our capacities to be

---

73. Smith, *Moral, Believing Animals*, 8.
74. MacIntyre, *Dependent Rational Animals*, 58.
75. Van Huyssteen, *Alone in the World?*, 276.
76. Bauckham, "First Steps," 230.
77. Conradie, *Ecological Christian Anthropology*, 117.

human and to manage God's planet arise from our solidarity with other creatures. As we have seen, science *does* provide support for human uniqueness. We *are* special. Therefore, since we are the only creatures that are both common and unique, ordinary and special, within his creation, God has placed us in charge of his earth.

We have understood ourselves for centuries as special beings, but we have not understood ourselves as animals. A key theme of this book is that these two ways of understanding ourselves must be held together. We are both fleshly and spiritual, biological and theological, earthly and heavenly. Grasping our animality helps us better understand just what freedom, love, and self-transcendence actually are. As Pascal tells us, we need to hold both our greatness and our baseness together, for this is the only way in which we can truly understand who we are and our place in the world.

4

The Bible I: God and Humans

THE BIBLE IS ADDRESSED TO HUMANS, not dolphins, chimps, or elephants. We are the species who needed it, who could understand it, and who could put it into practice—at least in part. As we saw in the last chapter, we are the only animals capable of self-transcendence and true moral life. We are the *ethical species* God has chosen and equipped for a special relationship with him; we are his managers and representatives in his earthly creation. In order to fulfill these roles, we need resources to guide us. God has provided us with two resources: the Bible and the person of Jesus Christ. In this chapter and the next, I shall look into the Bible and at Jesus in an attempt to glean some of the ideas that can provide that guidance.

Some Ideas on the Interpretation of Scripture

In chapter 1, I said that in Scripture, through stories, poems, songs, prophecies, letters, and so on, God reveals to us information about himself, about ourselves, about his other creatures, and about his world that helps us understand how we ought to live in his world. But discerning the meaning of Scripture and building a theology and ethics from it is a complex process. With respect to the Ecological Problem, there are Christians who disagree sharply about the meaning of various parts of the Bible. Of course, I am not going to resolve these controversies here, but in order to make my own approach to Scripture clear, I want make a few comments about its interpretation.

First, any interpretation should affirm that the Bible is, before all else, *the story of God*. It begins and ends with God. It is by God, about God, and for God. Like creation, it is God-centered—what we call *theocentric*. The Bible is also about his relationships—with himself (the Trinity), with humans, and with his entire creation. We often forget this. Our tendency to focus on ourselves and our modern cultural heritage of individualism and inwardness (discussed in chapter 2) lead us to see the Bible in terms of ourselves—as our story or the story of me and God. To be sure, Scripture has much to say to us personally, and that is important. But we must remember that in the grand scheme of the Bible, the main character is God, and God has concerns and purposes that go far beyond our individual, personal lives.

Second, we need to keep in mind the differences between the historical and cultural contexts of the biblical authors and our own. The people, places, and cultures of the Bible are far removed from those of today. For example, the Ecological Problem did not exist for the biblical authors in the way it does for us today. Questions about the troubled relationship between humans and the rest of creation were not foremost in their minds. But, as I noted in chapter 1, the Ecological Problem is, at its heart, a religious problem, and religious problems were prominent in the minds of the biblical authors. The Bible has a lot to say about these, so I think we will find rich resources in Scripture for our purposes.

Third, we must recognize that the Bible is not a book of science per se. By reading Scripture we will not learn about the second law of thermodynamics or the workings of the carbon cycle or the intricacies of open ocean food webs. God has revealed truth to us in two ways: (1) in what is called God's general revelation—the physical creation, the natural world; and (2) in what is called his special revelation—the Bible. I would add to these a third: God incarnate in Jesus Christ, a personal revelation of God. Humans can study these sources and learn about God and about what God's will might be for us and for his world. In general, Bible scholars and theologians study the Bible, and scientists study the natural world. Since God is the author of both, we assume that the information obtained from both sources will be complementary. It often is, but at times, it may be difficult to reconcile the findings of science with those of Scripture. This simply means that our science or our interpretation of Scripture or both are imperfect and incomplete. What we must keep in mind is that the Bible is not a book of modern science. The information it provides to us is of a different kind than that of science. It is a book of theology that addresses deeper questions about

God, existence, identity, meaning, values, goals, and destiny. These are questions to which modern science cannot speak.

Fourth, although the Bible is not a book about science, it needs to be interpreted in light of the information that science provides. As biologist Fred Van Dyke writes, "The Christian community must first commit to the concept that the best science . . . must be used and practiced as an aid to understanding the world and the Scriptures."[1] Christian neuropsychologist Malcolm Jeeves recalls some history that verifies this. In the early seventeenth century, the findings of Copernicus, Kepler, and Galileo showed that the earth was rotating on its own axis and that it was not the center of the universe but orbited the sun. These findings contradicted traditional understandings of cosmology, which had been used to interpret Scripture prior to this time. Jeeves continues, "Confronted with Galileo's challenges, our forbears went back to look again at those passages of scripture which, at the time, to them were the natural and irrefutable proof that the earth could not, as the astronomers were saying, be rotating on its own axis in space. The hard-learned lesson was that it is possible to take seriously what Scripture says but to be mistaken in interpreting it."[2] The findings of science changed our interpretation of Scripture. Take Psalm 96:10, for example, which says, "The world is firmly established." Instead of interpreting this literally as meaning that the earth is physically stationary with the universe rotating around it, we can interpret it metaphorically as signifying the Lord's faithful care and support of the earth.[3] In this way, we can improve our interpretations so they align more closely with the findings of science, and in so doing, we deepen our understanding of God and of his care for his earth.

Today, we may need to adjust our interpretations of Scripture in light of the findings of neuroscience, biology, and ecology just as the seventeenth-century interpreters changed theirs in light of the findings of astronomy and physics. The scientific evidence that we are ecophysical beings embedded within and dependent upon the ecosphere is overwhelming. We should take account of this as we approach Scripture. As New Testament scholar Douglas Moo notes: "The information Scripture gives us, while fundamental to everything else, is limited and

---

1. Van Dyke, "Bridging the Gap," 170.
2. Jeeves, "Changing Portraits," 23.
3. Ibid.

The Bible I: God and Humans

quite unspecific. Scripture must therefore be supplemented by what science tells us about the world God has made."[4] This can be a difficult and controversial task, and I will not resolve any of those difficulties here. But it is the approach I will try to follow.

Fifth, we must face the fact that none of us come to Scripture in a vacuum. We all come with a set of propensities, presuppositions, loyalties, a fund of knowledge, and a particular experience of the world. "All interpretations are, in various ways, products of a particular set of historical, ideological, theological and ethical convictions."[5] This is inescapable. For us flesh-and-blood humans, embedded as we are in our personal bodily histories, in our social systems, in our respective cultures, and in our local ecosystems, there is no such thing as a "view from nowhere." We all read the Bible from somewhere, as particular people, in a particular place, in a particular context.[6] In response to this reality, we can go to one extreme and say, "Oh well, I believe such and such, so this is how I will read Scripture, and I won't worry about other points of view, or about science, or about anything else"—a sort of "God said it, I believe it, that settles it" approach. Or, we can go to the other extreme and say, "Since everyone has a particular interpretation, and there is no 'objective' or 'true' interpretation, then Scripture can mean whatever I want it to mean, and so I can make up whatever theology I want and fit Scripture into that."[7] In my view, we must avoid both of these extremes. All Scripture is inspired by God (2 Tim 3:16) and contains truth in an absolute and ultimate sense, or what we might call God's Truth (capital *t*). But our capacity as fallible humans to access that Truth is partial and limited. As the Apostle Paul wrote, "Now we see only a reflection in a mirror . . . Now I know in part . . ." (1 Cor 13:12).

---

4. Moo, "Nature in the New Creation," 487.

5. Horrell, *Bible and the Environment*, 35.

6. Bouma-Prediger, *For the Beauty of the Earth*, 90. A good example of this is that in our modern culture, the Baconian idea of command-and-control over nature is central to the way we think about the world around us. If we do not have control over some aspect of nature, we are thinking of who or what has it and how we can get it. This mindset colors our interpretation of Scripture. We should not be surprised, for example, that we tend to approach the dominion mandate (Gen 1:26–28) in terms of the question of power and command-and-control. We have difficulty considering other possible approaches and meanings.

7. In reading the eco-theology of some liberal theologians, I have sometimes wondered if this is the approach they sometimes take.

Sixth, we should realize that although over time neither Scripture nor its doctrines change, the world and we humans change. As history unfolds, so also should the meanings and applications of biblical theology. As human culture changes, as new information becomes available, and as new problems arise, our approach to Scripture and its doctrines must take these into account. As theologian Loren Wilkinson says, "The very specificity of the Incarnation and the human nature of revelation testify to God's willingness to let the eternal message be expressed in local, temporal, limited terms . . . As time has unfolded, the content of the Gospel has unfolded. If we now see implications in Genesis, in the Psalms, in the Gospels, and in Paul's christological passages that are relevant to a theology of 'earthkeeping,' that is not to say that the Gospel changes, but that it unfolds new dimensions over time."[8] This unfolding must take place within the bounds of Christian orthodoxy and the checks and balances of the church community—past and present. But it must take place, or Christianity becomes static, hidebound, and irrelevant. In some cases, we find that a deeper understanding of existing doctrines or a revival of forgotten ones leads to the unfolding of theology and ethics in our changing contexts. For example, in what follows, we will find that a revival of the doctrine of creation would have a beneficial impact on our relationship to the rest of creation and its creatures.

I think that, as best we can, we must try to do two things, both as individuals and as a church. First, we should admit that we all come to Scripture with a particular perspective. We need to be reasonably and cautiously open to challenges to our perspective that are posed not only by others but also by science and by Scripture itself. This is a difficult, never-ending process, and to be sure we all do it imperfectly. But we need to try, and keep on trying. As Jesus said, "Ask and it will be given to you; seek and you will find; knock and the door will be opened to you" (Matt 7:7–8). Common sense and the community of the church are essential resources in this endeavor.[9] Second, we need to be prepared to change our perspective if the evidence warrants it. Like the Bereans who examined Scripture to evaluate what Paul was teaching them (Acts 17:11), we must think critically and be diligent and discerning about new ideas. (I hope you will do that with any new

---

8. Wilkinson, "New Age, New Consciousness," 27.
9. Fee and Stuart, *How to Read the Bible*, 14, 17, 26.

ideas you find in this book.) This is the only way, as far as I can see, that we deepen our understanding of reality and unfold our theology as our lives, cultures, and ecosystems change and unfold. We grow as Christians through learning about God's world (science), the study of Scripture, the ministry of the Holy Spirit, the teaching and fellowship of the church, and our efforts and practices in pursuit of Christ.

My approach to Scripture is, as I hope is clear by now, unequivocally physical and ecological. I take seriously the findings of neuroscience, biology, ecology, and the physical sciences that clearly show us to be eco-physical beings living in an eco-physical world. In my view, this approach will help us move closer to a "right" interpretation of Scripture—an understanding that meshes with what we find in his creation. I follow eco-theologian Steven Bouma-Prediger when he says, "My reading [of Scripture] . . . is unapologetically informed by ecology."[10] This is the approach I am taking, except that I am perhaps emphasizing the physical aspect of things more than he is. If you disagree, I hope you will bear with me and consider critically what I have to say.

Sixth and last, the goal of interpretation of Scripture should always be transformation, life, and action (2 Tim 3:16). Reading the Bible for reassurance or for comfort or because it's what Christians are supposed to do or because we enjoy it are all legitimate reasons. But Christianity is an inherently ethical faith, and as the Ecological Problem makes obvious and as our life experience demonstrates (if you are like me), we are all sinners—defective beings. We all need healing and transformation (Mark 2:17). Our ultimate goal, then, as we read Scripture should be life and action—that is, *ethics*—becoming the kind of people God wants us to be. As theologian Joseph Sittler put it, "The telos of doctrine is action, the fulfillment of right teaching is not right teaching but decision and deed."[11] Or as Bible scholars Fee and Stuart advise, "Read it, understand it, obey it."[12]

But this is without doubt the most difficult part of biblical interpretation. More often than not, we actually do understand what Scripture is saying, but we do not want to obey it. It doesn't "agree" with us (John 6:60). As some forthright Christian once confessed, "It is not the things in the Bible I don't understand that bother me, it's

10. Bouman-Prediger, *For the Beauty of the Earth*, 90.
11. Sittler, *Evocations of Grace*, 47.
12. Fee and Stuart, *How to Read the Bible*, 12.

the things I do understand." This, I think, is at the root of a lot of our interpretive controversies we have. We want to find some other way to interpret the Bible so we don't have to be and do what it says we should be and do. The problem is not interpretive technique but application (Matt 7:24–27).[13] But despite this, we should persist in the lifelong task of studying Scripture, formulating and reformulating our theology and ethics, praying, discussing it with others, and then seeking to put it into practice.

With these caveats in mind, I shall begin. My approach will be topical and theological. My goal is to try to interact with the whole of Scripture as best I can. Many passages will be cited along the way. I hope you will look up some of them and think critically about my interpretations. But I also hope you will think critically about your own perspective. We are all trying to get a little closer to God's Truth and to do a little better as his followers. I don't expect you to agree with me at every turn, but I hope you will engage humbly and honestly as we seek the heart and mind of God together.

God

As noted above, the Bible is about God, and God should be first in all things. So I will begin with God. As theologian Carl F. H. Henry writes, "The basic issue in ecology, as in every other human problem, is not only the nature of man, nor even the nature of nature, but ultimately also the nature and will of God."[14] In what follows I offer commentary on some aspects of God's nature that, I hope, will help us understand our place as humans in God's eco-physical world.

God Is a Unique and Transcendent Being—Wholly Different from His Creation

As Old Testament scholar Christopher Wright says, "The opening verse of the Bible [Gen 1:1] implies a fundamental ontological distinction between God as creator and everything else as created . . . This duality

---

13. For a revealing account of this phenomenon with respect to Jesus' teachings on wealth, see Metzger, *Consumption*, 1–15.
14. Henry, *God, Revelation, and Authority*, 100.

between the creator and the created is essential to all biblical thought and to a Christian worldview."[15] Thus far I have been critical of dualisms, such as body-soul and human-nature dualism, but I propose that the dualism between God and his creation is biblical and that we should embrace it. As Wright says, it is fundamental to a Christian worldview. This God-creation dualism is sometimes called God's *transcendence*. God's existence is radically other than and transcendent over all other existence; he is above and beyond all creation and all creatures, including us: "For I am God, and not a man—the Holy One among you" (Hos 11:9); "'For my thoughts are not your thoughts, neither are your ways my ways,' declares the Lord. 'As the heavens are higher than the earth, so are my ways higher than your ways and my thoughts than your thoughts'" (Isa 55:8–9). God is limitless, all-powerful, perfect, certain, and omniscient. His creation and we humans are none of these things. Although God is a radically different being from us, he has revealed himself to us in Scripture and in Jesus Christ—the human incarnation of God. We humans can grasp these things in limited ways. I shall address God's self-revelation in Christ near the end of the chapter.

The doctrine of God's uniqueness and transcendence has two important implications for my purposes here. First, it means that nature is not divine. Pantheism, the belief that all things are one and the same with God, is rejected. Nature is not divine and is not to be worshiped in any form (Exod 20:4–5). Second, we humans are not gods. We are creatures—part of creation. Both of these implications must be kept in mind in order to have a proper understanding of our human relationships with the rest of creation and with God. Neither we, nor nature, nor anything in it can be equated with God.

God the Creator of All Things

"In the beginning God created the heavens and the earth" (Gen 1:1).[16] So begins the great narrative of Scripture. Bible scholar Tremper Longman says that this is the "single most important theme" of the opening

---

15. Wright, "'Earth Is the Lord's,'" 221. What Wright means is that God is a completely different kind of being from creation (including us humans). God and creation are totally different things.

16. See also Neh 9:6; Ps 8:3; 146:6; Prov 3:19–20; Isa 45:18; 66:1–2; Jer 10:11–13; 27:5; John 1:3, 10; 1 Cor 8:6; Col 1:16; Heb 1:10; 11:3; Rev 4:11.

chapters of Genesis: "Yahweh [the Lord] created the cosmos!"[17] It is hard to overstate the importance of this. As Christians we are, or should be, *creationists*. That is, we *see and understand* the world and all things in it as products of the intentional creative action of God. Whether we hold to an interpretation that God created the universe in six literal days or to another view is beside the point. As Longman says, the key idea is that God is responsible for the existence of everything—the sky, the land, the seas, the creatures, the ecosystems. These are all the handiwork of God's intentional creative action.

We modern Christians have largely forgotten the doctrine of creation, if not in word, then certainly in life and action. In the modern period (and perhaps earlier), Christianity lost "an effective doctrine of creation such that the human relationship to other creatures as fellow creatures gave way to an exclusively vertical relationship of humans to nature."[18] That is to say, in our modern worldview, as a result of the things we covered in chapter 2, such as philosophical mechanism, command-and-control instrumentalism, Descartes' human-nature dualism, and the modern turn to the self, we no longer view the world as created by God, nor do we see its creatures as our fellows. We live within the paradigm of modern materialistic, self-interested, economic utilitarianism rather than within the paradigm of biblical creationism. We see ourselves as on top of creation rather than within it, and we regard created beings as objects for our use rather than as fellow creatures of God, the handiwork of his intentional creative act.

Peter Harris and his wife, Miranda, are the founders of A Rocha, an international Christian organization promoting ecological care. Harris emphasizes the truth that we Christians do not follow the biblical view of the world as created:

> From beginning to end, the Bible affirms that God is the Creator. This, and not the human condition, is the true starting point for both understanding and caring for the world around us. We discover who we are and what the world is in relationship to God, rather than the other way around. Nevertheless, this understanding of God is at odds with the conviction that achieving human happiness is at the heart of the human enterprise, and thus of our environmental relationships. Assertions that belief

17. Longman, *How to Read Genesis*, 79.
18. Bauckham, *God and the Crisis of Freedom*, 165.

> in the Creator is not primarily an affirmation about humankind comes as a big shock to most people, including Christians. We have to recognize that we are not saying, "I believe in God who made me" but, as the Apostles' Creed states, "I believe in God, maker of heaven and earth."[19]

As Harris points out, in thought, word, and deed, whether Christian or not, we modern humans tend to see the world, not in relation to God but in relation to ourselves, not in terms of God but in terms of ourselves. Thus we see the creation as *about us*, as there *for us*—for our needs and wants. We think of nature in a utilitarian way—as a collection of resources for us to use so that we can pursue our personal, social, and economic goals of enjoyment, happiness, security, and fulfillment.[20] To use C. S. Lewis's term, we tend to see nature as "pure datum."[21] Indeed, as Lewis points out, if we did truly believe in God the creator of all things, we would find ourselves put in our place alongside whales, lions, and cattle, our fellow creatures.[22] Peter Harris goes on, "A biblical perspective is that 'for [and] in [Jesus] all things in heaven and on earth were created, things visible and invisible' (Col. 1:16a, NRSV). Here is a theology that frames everything we know in time and space as *God's artifacts, the work of his hands* . . . Thus we live in 'creation,' not 'the environment' or 'nature,' and we ourselves are in a different relationship with those around us, sharing a common created humanity."[23]

The doctrine of God as Creator is potentially transformative. If we repented of our self-centered, self-referential, secular, economic, utilitarian view and embraced God as the creator of all things, we would treat his creatures and his ecosphere very differently from the way we do now. "If we believe that the earth is God's creation, then 'we would act less recklessly . . . , not only in irreversible ecological affairs, but in quieter relationships with the earth and its creatures day to day.'"[24] We would be grateful people, continually thanking God for every creature we encountered, from insects to soaring eagles. Similar to Saint Francis, we would be more humble people, reluctant to exploit and destroy

---

19. Harris, "Environmental Concern," 11.
20. Berry, *Unsettling*, 56.
21. Lewis, *Reflections*, 83.
22. Ibid., 84.
23. Harris, "Environmental Concern," 12–13; italics mine.
24. Douglas, *Wilderness Sojourn*, 30.

Dusty Earthlings

God's work, but eager to respect, preserve, and care for it (Gen 2:15). We would be more joyful people, continually rejoicing in the wonders of God's handiwork. And we would all be students of God's creation—ever interested and eager to learn about it.

We need to broaden our understanding of God. As J. B. Phillips noted in his book *Your God Is Too Small*, we have a conception of God that is far too small.[25] Theologian Joseph Sittler understood this back in 1975: "We are not just *homo sapiens* crawling across the surface of the world. We are of the world. We are produced by it. Therefore, any notion of God which is to come to us as a private word whispered into our individual ear—'Buck up, old boy, I'm on your side'—this is not enough. The only salvatory God must be a God who has the whole wide world in his hands. If he is not the God of everything, he is not an adequate God for my anything, no matter what that anything might be."[26] Our modern God is indeed too small. A God whom we can tuck in our back pocket to be pulled out when we need him is no God at all and no savior. God is not our sidekick, waiting to be called on to help us when we're down, but otherwise ignored. God is God, infinitely greater than we are—Creator and Lord (1 Cor 12:3). Given our current self-referential theology, attitudes, habits, practices, culture, and economy, grasping and worshiping God as God the Creator is a big challenge.

God the Owner of Everything

Chris Wright notes, "The whole earth belongs to Jesus by right of creation, by right of redemption, and by right of future inheritance—as Paul affirms in the magnificent cosmic declaration of Colossians 1:15–20. So wherever we go in his name, we are walking on his property."[27] In the United States we have a powerful tradition of private property. It is written into our laws, and our culture contains strongly held values of private ownership and personal sovereignty over what we own. As cultural heirs of a strong form of individualism and the command-and-control attitude toward the world, we are attracted to the power of ownership. I understand that there are other cultures where private ownership is not

---

25. Phillips, *Your God Is Too Small*, 8.
26. Sittler, *Evocations of Grace*, 215.
27. Wright, "'Earth is the Lord's,'" 222–23.

so strong—where people have a sense of common ownership among families, clans, or tribes, or even no ownership at all.[28] But whatever may be the views and customs of the culture in which we find ourselves, for us Christians, the Bible presents us with the first principle that, before all else, *everything belongs to God*.[29]

In the Old Testament, land is understood not as a possession but as a *gift* to be held in trust by humans.[30] As Abraham traveled through the land of Canaan, God said to him, "To your offspring I will give this land" (Gen 12:7). This promise is repeated several times to Abraham, his descendants, and later to Moses and the Israelites.[31] It contains the a priori assumption that God owns the land and therefore can make a gift of it to whomever he chooses, in the way he chooses. When the Israelites finally do receive the land and occupy it, land ownership is not based on natural law, economic exchange, or power, as it is in our modern culture, but on a *land-gift-trust* theology.[32] "Land holdings were the allotments of the divine giver, and therefore were held in trust from God."[33] In the book of Leviticus, the Lord instructs the people that "the land must not be sold permanently, because the land is mine and you are but aliens and my tenants" (Lev 25:23). Although the various families and clans of Israel "owned" allotments of land, God retained first title to it. The land was part of the covenant. Thus, Israel's continued occupancy of it was contingent on their obedience to the covenant (Lev 18:28).

This idea is illustrated in the story of King Ahab and Naboth (1 Kgs 21:1–29). Ahab wanted Naboth's land, but despite Ahab's offer of better land and a good price (this would mean a great deal in modern economic terms), Naboth would not accept because he saw the land as belonging to God. It was not his to "give, sell, or exchange."[34] In the end Naboth's commitment to the principle that God, not he, was the

---

28. As part of the rule for his order, St. Francis of Assisi forbade any ownership at all. Friars had no property.

29. Exod 9:29; Lev 25:23; Deut 10:14; 1 Chr 29:11; Job 41:11; Pss 24:1; 50:10–12; 95:3–5; Matt 5:34.

30. Brueggemann, *Land*, 45.

31. Gen 13:15, 17; 15:18; 17:8; 24:7; 26:3–4; 28:13; 35:12; 48:4; 50:24; Exod 13:5, 11; 32:13; 33:1; Num 11:12; Deut 1:8; 2:31; 9:5; 11:9; 34:4.

32. Wright, *Old Testament Ethics*, 90.

33. Ibid.

34. Ibid.

ultimate owner of the land cost him not only a lucrative land deal but also his life. Similar violations of this land-gift-trust theology were criticized by the prophets (Isa 5:8; Mic 2:2). Since God owned the land, he demanded equity, justice, care, and accountability in how it was used and managed.[35]

I suggest that under the new covenant of Christ, God's ownership of the land is not abrogated (Matt 5:17). Just as God remains Creator, he also remains Owner of creation. In the perspective of the New Testament, Jesus is Lord of the earth (Matt 28:18; 1 Cor 12:3; Rev 1:8). By virtue of God's creation and redemption of all things, he is owner, not simply of the ancient land of Israel, but of the entire earth—the land, the sea, the atmosphere, the creatures, the ecosystems, the resources—everything, even us (1 Cor 6:20, 7:23).[36] Theologian James Nash points to this when he characterizes "ecological sin" as our "acting like the owner of creation with absolute property rights."[37] As worshipers of God in Jesus Christ, our attitude and approach toward land and all it contains ought to be that in a primary sense Jesus owns it all. We have no "absolute property rights" to anything. And as Chris Wright points out, the corollary to this is that all things are received gratefully as divine gifts. For us Christians, to "own" land (or anything) is to receive it and hold it in trust before God, similar to the way the ancient Israelites were supposed to have done. Our theology of divine ownership and land as gift-held-in-trust should supersede any human system of property ownership or rights. Christians may "own" land in a legal, economic, or social sense, but we ought to always use, manage, and care for it (and the creatures it contains) in a way that recognizes the God whom we worship as its primary and ultimate owner. We are merely trustees, caretakers, accountable to God for how we treat the land that God has allowed us to "own" for a brief time. Like ecosystems, watersheds, and the routes of migratory animals, this Christian land ethic should transcend all national, political, tribal, and ethnic boundaries.

Divine ownership of all things, like the doctrine of God as Creator, is in conflict with our modern norms of ownership, land use, economic utility, self-interest, and individual freedom. We are so deeply saturated with these ideas that it is extremely difficult for us to grasp the notion

---

35. Ibid., 95.
36. Wright, "'Earth is the Lord's,'" 222–3.
37. Nash, *Loving Nature*, 119.

of divine ownership, let alone live it out. But how *do* we live it out? There is a great need for a Christian land-earth ethic.[38] Christian theologians and ethicists should develop this and provide us laypeople with practical ways in which we can apply this to our lives and property. If theologians did this, and we Christians lived it out, then it would perhaps improve our capacity to address the Ecological Problem, and give powerful testimony of the reality of God to the world.

God the Provider

God is not only creator and owner, he is also the provider of all of creation.[39] In and through Christ, the universe is supported in its existence (Col 1:17; Heb 1:3).[40] Psalm 104 is, perhaps, the prototypical text that describes God's providential care of his world. He provides for the trees, which in turn are places for the birds to make their nests (vv. 16–17). The mountains are the "property" of wild goats, and the cliffs are the home of the hyrax (v. 18).[41] God also maintains the rhythms and cycles of nature: "The moon to mark off the seasons, and the sun knows when to go down" (v. 19). Psalm 104 sees God's providence and care as *holistic*. It is for all of his creation and all his creatures. As in Genesis 1, we see a picture of creation as an integrated whole. The psalm also contains the idea that God provides for the nonhuman creatures for their own sake, not simply for human use. Springs pour water into the valleys to give water to the grazing animals of the field and to provide a place for the birds to nest and sing (v. 10). The trees are for the birds, and the high mountains and cliffs are for the wild goat. Humans are included, to be sure, but they are included as "one of many kinds of living creatures,"[42] part of the larger whole.[43]

---

38. Bouma-Prediger, "Living on the Land," 5; Baer, "Land Misuse," 1241.

39. Pss 36:6; 65:9–13; 89; 104; 145:15–16; 147:8–9, 16–17; Matt 6:26–28; Luke 12:24–28; Acts 17:25, 28.

40. Wilkinson, "Cosmic Christology," 29.

41. The hyrax is a rabbit-sized mammal that looks something like a guinea pig. It inhabits rocky areas in Africa and the Middle East and is sometimes called a rock badger or coney.

42. Bauckham, *Living with Other Creatures*, 10.

43. Limburg, *Psalms*, 354.

Again, in our modern view, we tend to see God's providence in terms of ourselves. We focus on God's care for us, and it is hard for us to think of God caring for other creatures or for the whole creation and even harder for us to live in a way that recognizes these theological (and ecological) realities. As theologian Bernard Anderson says, "The doctrine of creation affirms that every creature is assigned a place in God's plan in order that it may perform its appointed role in serving and glorifying the Creator."[44] As God's ethical species, his representatives in his world, we humans ought to seek not only our proper place within the whole, but we ought also to live in a way that reflects God's care for other creatures and for the whole of his beautiful planet.

Finally, understanding God as the great provider should help us trust him for our own needs. In our modern world, we have been caught up in an anxious drive not just to meet our needs but to insure our ever-increasing affluence.[45] Embracing our dependent creaturehood should help us renew our faith in God's providence for us, find contentment in what we have, and give up our anxious pursuit of more and more (Matt 6:24–34).[46]

God's Immanence: His Direct Relationship with His Creation and His Creatures

As we noted previously, God is a different kind of being from his creation, but this does not mean that he has no contact with it. God indwells his creation and relates directly to his creatures.[47] This is what is called God's *immanence*, and it must be kept in balance with his *transcendence*. This is a biblical idea that we really have trouble with—the notion that, apart from us humans, God is directly involved with his creatures and his creation. We tend to think of God as strictly related to us personally or to our human social group, or we think of God as spiritual, separate from the physical world. For reasons we have already discussed, we modern people have tended to remove God from the natural world and to think of the natural world as mere stuff, the realm

---

44. Anderson, "Earth Is the Lord's," 14.
45. Bauckham, *Bible and Ecology*, 74.
46. Bauckham, *Living with Other Creatures*, 142.
47. Ps 139:7–12; Jer 23:24; Acts 17:28; Eph 1:23; 4:6, 10.

The Bible I: God and Humans

of economics and science. But "nature and God are not as separate as we usually think."[48] God speaks to Job out of a whirlwind, and uses Balaam's donkey to communicate with him (Num 22:21–34). When the disciples are caught in a storm on the Sea of Galilee, Jesus rebukes the wind and waves, and it becomes calm (Mark 4:39). God is present with and among his creatures. Because of God's immanence within the creation, "we should have an appreciation for all that God has created."[49]

A friend of mine once drove from San Diego to Colorado, a journey that passes through some of the vast western deserts and plains. As he told me about his trip he spoke of driving through areas that he described as "God-forsaken wasteland, miles and miles of nothing." We sometimes hear people speak in similar ways about undeveloped land as places where there is "nothing." Seeing the world in terms of ourselves and in terms of economic utility, we evaluate regions of the earth that are inhospitable to us humans or that do not seem attractive or useful as "God-forsaken," empty, and worthless.[50] The Bible acknowledges that some lands are less desirable from a human perspective,[51] but in the overall scheme of creation, the value of landscapes and ecosystems is not simply about us humans. It's about God. From a divine perspective, land less hospitable to humans may be valuable and "useful." My friend misunderstands God's relationship with his creation. Deserts, steppes, and ocean depths are not "God-forsaken," nor are they "nothing." In truth, they are complex, living communities of microbes, plants, and animals who have astounding capacities to survive and thrive under incredibly difficult conditions. They are good in God's sight (Gen 1:31), and he provides for them (Ps 104). If we took time to stop and learn about some of these places and the creatures that God has placed in them, we would be filled with awe and wonder. God indwells all places in his creation and cares for the creatures who live there. He values

---

48. Erickson, *Christian Theology*, 329–30.

49. Ibid., 337.

50. This may, in part, reflect John Locke's labor theory of value, which has been incorporated into our American legal and economic systems. Land is valued in proportion to the amount of human labor put into it. Thus land that has been modified by humans (cleared, built on, fenced, etc.) is more valuable than land that is not modified by humans (Locke, *Two Treatises*, 2.27).

51. Ps 1:3; Isa 41:17–20; Jer 17:8; Joel 3:18–19.

them—even if we humans do not. As we shall see, this is part of God's message when he speaks to Job (Job 38–41).[52]

Old Testament scholar Terence Fretheim warns that "to remove God from intimate involvement with the world invites the introduction of other divine powers at the earthly level (Mother Nature?). Or, more drastically, the radical transcendence of God may lead to the death of God in the life of the world."[53] In other words, when our worldview excludes God from his physical creation, we invite other "idols" to take his place. The world may become Gaia, or the Earth Goddess, or some other pseudo-deity. But far more commonly, we replace God with ourselves. *We* become the pseudo-deity—the god of creation—either as individuals or as modern human society, and the rest of creation becomes mere stuff for our use. This idea is part of our American culture, so, like other cultural ideas, we tend to absorb it unconsciously and live it without being aware of it. But as our abuse of the land and seas shows, we humans make poor gods of the earth. It is God who indwells his creation as its Lord and God—not us.

God Delights in His Creation[54]

Throughout the creation account of Genesis 1, we read the refrain "And God saw that [what he created] was good" (vv. 3, 10, 12, 18, 21, 25). Finally, when all is accomplished, God contemplates all that he has made and finds it "very good" (Gen 1:31). The word *good* could be translated "lovely, pleasing, beautiful."[55] In other words, God takes joy and satisfaction in his creation. It is a source of pleasure and delight for him. Since God delights in his creation, so should we. In imitation of God (in the image of God), as a matter of practice, we ought to find joy and pleasure in God's creation and his creatures (more about this in chapter 7).

---

52. Deserts can also be places of spiritual retreat and renewal. See Douglas, *Wilderness Sojourn*.

53. Fretheim, "Nature's Praise," 17.

54. Hoezee, *Remember Creation*, 21.

55. Brueggemann, *Genesis*, 37.

The God of Kenosis

The word *kenosis* is derived from the Greek word *kenoō*, which denotes an act of self-emptying, of voluntary self-abasement—the giving up of position, power, and privilege.[56] In the Bible it appears in chapter 2 of Paul's letter to the Philippians:

> [Christ], being in very nature God,
> did not consider equality with God something to be grasped;
> but made himself nothing,
> taking the very nature of a servant,
> being made in human likeness.
> And being found in appearance as a man,
> he humbled himself and became obedient to death—
> even death on a cross! (Phil 2:6–8).

Christ "made himself nothing" here translates the word *kenoō*. Thus, Jesus Christ, God the Son, in his incarnation, passion, and death, emptied himself of his divine position, power, and privilege—a divine act of self-denial, voluntary self-limitation, and sacrificial love. In this passage, Paul exhorts us to imitate Christ: "Your attitude should be the same as that of Christ Jesus" (Phil 2:5). Thus the kenosis of God in Christ has direct ethical implications for us in how we live and act in this world. I shall develop this further in chapter 7, but my point here is that the redemptive act of God in Jesus Christ means that kenosis—loving, voluntary self-denial—is a fundamental aspect of God's character and action.

For the sake of who or what did Christ empty himself? The traditional answer is, for the sake of humanity (or human souls), but this is too narrow. Christ's kenotic act of redemption is cosmic in scope. Scripture makes clear the pervasive lordship of Christ over all things.[57] Since the lordship of Christ is cosmic, so also is his redemption: "For God was pleased to have all his fullness dwell in him, and through him to reconcile to himself all things, whether things on earth or things in heaven, by making peace through his blood shed on the cross" (Col 1:19–20). As the ethical species among God's creatures and as the main source of sin in the world, we humans are the principal object of God's redeeming grace, but we are not the only object. Beyond us, Christ's

---

56. Bauer et al., BDAG, 539.
57. Matt 28:18; Rom 11:36; 1 Cor 15:27; Eph 1:20–23; Phil 2:10–11; Col. 2:9.

kenotic act of redemption includes all of creation. The gospel of Jesus Christ is (or ought to be) good news not just for humans but for all creatures and all creation (Col 1:23).

The kenotic character of God is manifested also in his act of creation. God who is infinitely abundant voluntarily gave up something of himself in order to create something other than himself—the creation. He emptied himself in order to provide a "space" in which the creation could exist.[58] Thus, the very act of divine creation was an act of kenosis. As theologian Jürgen Moltmann puts it, "The created world does not exist in 'the absolute space' of the divine Being; it exists in the space God yielded up for it through his creative resolve. The world does not exist in itself. It exists in 'the ceded space' of God's world-presence."[59] Divine creation was divine kenosis.

Also, God's ongoing relationship with the world is kenotic in nature. Again Jürgen Moltmann helps us understand that God's "kenotic self-surrender"[60] is manifested not only in creation and incarnation but also in God's love for the world:

> A love that gives the beloved space, allows them time, and expects and demands of them freedom is the power of lovers who can withdraw in order to allow the beloved to grow and to become. Consequently it is not just self-giving that belongs to creative love; it is self-limitation too; not only affection, but respect for the unique nature of the others as well . . . It is not God's power that is almighty. What is almighty is his love . . . God acts in the history of nature and human beings through his patient and silent presence, by way of which he gives those he has created space to unfold, time to develop, and power for their own movement.[61]

In other words, the way God shows his love for us humans (and all his creatures) is by self-limitation—withdrawing, allowing us the space to

---

58. I am thinking metaphorically here, not in terms of literal space-time within the created universe as we know it. God, of course, is outside of, or above, space-time. I am suggesting that the superabundant nature of God's existence required that he somehow limit his superabundance in order to create the universe as something other than himself. In effect, God's act of creation included a yielding up, a self-emptying, a kenosis of himself.

59. Moltmann, *God in Creation*, 88, 156.

60. Moltmann, "God's Kenosis," 141.

61. Ibid., 147–49.

"unfold" and "develop" as his creatures. God loves us and his creation by "letting-be, by making room, and by withdrawing himself,"[62] affording us and all his creatures the freedom to become what he created us to be.

In his first letter, the Apostle John writes, "God is love" (1 John 4:8, 16). Indeed, we have seen that in Christ's kenotic act of redemption, in God's great act of creation, and in God's continuing care of his creation, he is love in that he voluntarily gives of himself and limits himself so that his creatures may flourish. God is a kenotic God of love.

Conclusion: God is the King of Creation

In sum, God is a being who is radically different from his creation, including us humans who are one of his creatures. He is the creator, owner, and provider of all things. He is both transcendent to and immanent within his creation. He delights in his creation, and he is a God of kenosis, of voluntary self-emptying and self-limitation by which he has created, supported, and redeemed his world. As theologian Steven Bouma-Prediger writes, "The biblical witness confirms what those most famous of photographs [of earth] from space portray—that the blue-green sphere on which we live is finite. Only God is unlimited in power, knowledge, duration, presence, and compassion."[63] God is *the* king of his creation.[64] We humans are not *the* king. As part of creation, we are subjects of God's kingly rule manifested in his loving creative, providential, and redemptive care.

Humans: Us

The prophet Ezekiel spoke an oracle against the king of Tyre: "In the pride of your heart you say, 'I am a god; I sit on the throne of a god, in the heart of the seas.' But you are a man and not a god, though you think you are as wise as a god" (Ezek 28:2). This passage gives rebuttal to Giordano Bruno's grandiose pronouncements that I cited at the end of chapter 2. I also noted there that Bruno's comments reflect the goal of modern humans to become masters of the world—to achieve

---

62. Moltmann, *God in Creation*, 88.
63. Bouma-Prediger, "Creation Care and Character."
64. Deut 6:4; Ps 47; Isa 45:5; Matt 28:18; Rom 11:36; Eph 1:20–23; Phil 2:10–11.

separation from and dominance over creation and to become, in effect, gods. This lies at the heart of the Ecological Problem, and it conflicts with the biblical view. *We are not God. We are creatures* (Eccl 5:2). This biblical truth cannot be overemphasized. It has been noted by numerous theologians and scholars.[65] Douglas John Hall writes, "Creaturehood is the core of Christian anthropology."[66] Chris Wright declares, "We are animals among animals. We are creatures, earth-creatures—*ādām* from the *ădāmâ* (or humans from the humus, or soil)."[67] Preacher Elizabeth Achtemeier says, "We share a material, bodily existence with all living creatures, and we are not some sort of unique spiritual beings elevated above the rest of nature."[68] Old Testament scholar Daniel Block points to the creation narratives (Gen 1–2) to emphasize that we "human beings are earthlings."[69] He gives six reasons for this:

> (1) They [Humans] are one with the land animals, which were also created on day six (Gen 1:24–31). (2) They share the divine blessing and the mandate to multiply and fill the earth (Gen 1:22, 28). (3) They share vegetation as food with the animals (Gen 1:29–30). (4) The appellation for the human species, *ādām*, "humankind," derives from the same root as *ădāmâ*, the Hebrew word for "ground/land" (Gen 1:25; 2:5, 6). (5) Since they are made of "dust from the ground" (Gen 2:5–7), like the animals . . . human beings originate in the earth itself (Gen 2:7). (6) They share with animals the breath of life and the generic classification as a *nepeš hayyâ*, "living creature" (Gen 2:7; cf. Gen 1:20, 21, 24). As earthlings, human beings participate in the covenant that God made with the earth and its living things.[70]

Block notes, however, that humans are "not simply one among equals," calling us the "climax of the creation week."[71] While it is true that we have been gifted with unique capacities and with the job of managing God's earthly creation, Block is mistaken in calling us humans the "climax" of the creation narrative. Theologian Jürgen Moltmann correctly recenters the meaning of the text where it belongs—on God. The climax

---

65. See, for example, McClendon, *Ethics*, 59.
66. Hall, *Professing the Faith*, 337.
67. Wright, "'Earth Is the Lord's,'" 225.
68. Achtemeier, *Nature, God, and Pulpit*, 52.
69. Block, "To Serve and to Keep," 126.
70. Ibid.
71. Ibid.

of creation is not the creation of humans on the sixth day but God's *Sabbath* on the seventh. On that day, God rests and contemplates his creation,[72] finding it "very good" (Gen 1:31). I shall discuss these ideas further presently.

But despite Scripture's clarity that we are dusty earthlings, "the theme that man, sinful though he be, occupies a position on earth comparable to that of God in the universe, as a personal possession, a realm of stewardship, has been one of the key ideas in the religious and philosophical thought of Western civilization regarding man's place in nature."[73] The idea that humans are made "in the image of God" is, of course, contained in Scripture,[74] but we must be mindful of our persistent inclination to think more highly of ourselves than we ought to think (Rom 12:3). Our constant tendency toward pride and self-absorption influences our interpretation of Scripture, distorts our view of ourselves and our relationship to the rest of creation, and puts us in danger of violating the first commandment (Exod 20:3).

Theologian Wolfhart Pannenberg notes that the idea of humans having an immortal soul "was conceived . . . in biblical and Christian terms as a *supraterrestrial* distinguishing mark and dignity that elevates humanity *above the entire cosmos* and sets it at God's side over against the cosmos."[75] While the idea of humans having (or being) immortal souls might be a basis for ascribing to them honor and dignity, it must not abrogate our creaturehood, nor elevate us "above the cosmos," nor set us "at God's side." This, of course, is idolatrous. We are not seated at God's side; Jesus is (Col 3:1). *We are neither gods nor demi-gods.*[76] *We are dusty earthlings*. As we modern humans face the Ecological Problem in this twenty-first century of creation's existence after Christ's first advent, this fundamental, theological and scientific truth must be kept ever before our minds and hearts.

We should interpret Genesis 1:26–28 in light of the overwhelming evidence from science and from ordinary experience that we are eco-physical beings embedded within the eco-physical world. As we shall see in the next chapter, if we are "in the image of God," we are so in a

---

72. Moltmann, *God in Creation*, 277.
73. Glacken, *Traces*, 155.
74. Gen 1:26–27; 5:1; Ps 8:5.
75. Pannenberg, *Anthropology*, 27; italics mine.
76. Bauckham, *Bible and Ecology*, 19, 27.

physical way, in a biological way, in an earthly way, in an ecological way, in a creaturely way, but not in a divine way. To be sure, our likeness to God carries ethical meaning, but it does not carry ontological meaning. "We are part of the ecosystem, not of the heavenly court."[77] God and only God is above and outside of creation. We dwell below and within it. Several theologians have issued prophetic warnings about the danger of confusing ourselves with God. Douglas John Hall observes that it may be harder for us humans to confess our creaturehood than it is for us to confess our sin.[78] Australian Ernst Conradie notes that "the difference between Creator and creature is crucial if we are to maintain respect and reverence for the Creator and to acknowledge the humility of the creature."[79] British theologian Colin Gunton writes, "We belong with and alongside God's other creatures, irrespective of the distinct qualities which the relationship between God and humans may have. We have more in common with the 'lowest' of the creatures than with God."[80] Bernard Anderson comments that we humans "cannot escape" who we are: creatures who, like all creatures, are "dependent upon the sovereign judgment and grace of God."[81] Joseph Sittler sums up:

> Man is formed by God. Man *is* because God is; God wills man to be. God is the Creator; man is created creature ... God makes man out of, within, and absolutely dependent upon the *whole* world that he has made. I can live without food for a month, without water for a week, without air for perhaps six minutes. I am stuck with God, stuck with my neighbor, and stuck with nature (the 'garden'), within which and out of the stuff of which I am made. I may love God, hate God, ignore God. But I can't get unstuck from God. I may love my neighbor, hate my neighbor, or ignore my neighbor. But I can't get unstuck from my neighbor. And I may love the world, hate the world, or try to ignore the world. But I cannot get unstuck from the world.[82]

We are created, finite beings, radically different from the Creator, and, as Gunton says, much more like other creatures than like God. This is

77. Towner, "Future of Nature," 33.

78. Hall, *Professing the Faith*, 268. As we will see in the next chapter, these two things are connected.

79. Conradie, *Ecological Christian Anthropology*, 53.

80. Gunton, *Triune Creator*, 205.

81. Anderson, "Earth Is the Lord's," 10.

82. Sittler, *Evocations of Grace*, 203–4.

essential to a proper biblical worldview and to correctly understanding who and what we are and our role in God's world.

Physicality

As I proposed in chapter 1, whatever view of human nature you may hold, you should agree that we *are* physical beings. To exist as a human is to be a physical body. The prototypical text affirming this is Genesis 2:7: "Then the Lord God formed a man from the dust of the ground and breathed into his nostrils the breath of life, and the man became a living being."[83] Similarly, Eve was formed later from the man's rib (Gen 2:21–22), again affirming the materiality of both male and female together. Lawson Stone says that "when Genesis 2:7 speaks of Adam's being formed from the dust of the ground, it stresses his fragility and commonality with the animals. The origin of Adam's body from the dust of the ground comports well with an affirmation of humanity's basic continuity with the material order."[84] Stone rejects claims of a "special creation" of humans over other animals. "Adam does not differ from the 'beasts of the field.'"[85] The term *nefesh hayyah* or "living being" is applied not just to humans but to all of animal life (Gen 1:20, 24).[86] Chris Wright agrees: "The conclusion of this verse, 'the man became a living being,' uses the same word (*nepeš*)[87] repeatedly used of all the other living beings"[88] "Our essential nature as one species of animal among the rest is highlighted, even after the splendid words of our creation, by the fact that God provides the same food for us as for the rest of the animals" (Gen 1:29–30).[89] The Hebrew Bible views a human being as an "integrated whole"[90] made of dust, the common material of all creation.

In the first three chapters of Genesis, humans are pictured as existing within the ecological context of a living community. Genesis 1

---

83. Numerous other passages affirm our earthly dustiness: Gen 3:19; 18:27; 1 Kgs 16:2; Job 30:19; 34:14–15; Pss 90:3; 104:29; Eccl 3:20; Isa 40:15; 1 Cor 15:47–49.

84. Stone, "Soul," 50.

85. Ibid.

86. Ibid., 51.

87. *Nepeš* is a different way of writing *nefesh*.

88. Wright, *Old Testament Ethics*, 117–18.

89. Ibid., 118.

90. Green, "'Bodies—That Is, Human Lives,'" 158.

places us within the holistic order of the cosmos. Genesis 2 places us within the garden of Eden, a forest ecosystem formed by the hand of God, a specific location identified by the four rivers (Gen 2:10–14).[91] A bumper sticker I once saw suggested that God's original idea for human life was to hang out in a garden with naked vegetarians. Joking aside, it seems there is little question that the creation story of Genesis sees the primary state of humans as organic beings embedded *within* the ecophysical world, not outside of it.

New Testament scholar Joel Green is one of a few scholars who have engaged with Scripture in light of the findings of science, specifically neuroscience. He sees all of Scripture, including the New Testament, as affirming the physical nature of humans. The Bible claims "the embodiment of humanity in relationship to all creation, the full integration of the human being within the human family and, indeed, within the whole cosmos."[92] Green notes that the Apostle Peter emphasized our bodily existence. Peter's use of the Greek word *soma* (body) in 1 Peter 2:24 in reference to Christ bearing our sins "in his body on the cross," is a "profound affirmation of bodily existence and of the significance of embodied, human suffering . . . [Peter's] emphasis on embodied existence provides life in this world its fullest significance and it serves as the basis for his emphasis on a faithful manner of living in the material world. Human physicality also ties Peter's audience to the rest of creation, thus pressing the question how their suffering participates in the situation of the cosmos, and, perhaps more to the point, how their liberation is tied to the fate of the cosmos. Importantly, the work of Christ in death and exaltation has repercussions for humans and for the cosmos."[93]

Similarly, New Testament scholar James Dunn remarks on Paul's use of the word *soma* in his epistles.[94] He suggests that a better

---

91. I am speaking conceptually here. Whether you interpret the garden's location literally or figuratively, the point is that the narrative sees it as located in the earth. Humans in their primordial state are set in a *place*. This suggests that "placeness"—possessing, belonging to, and living in a particular place—may be a fundamental biblical feature of our human creaturehood. If that is the case, then our modern rootlessness may help explain the Ecological Problem.

92. Green, "'Bodies—That Is, Human Lives,'" 152.

93. Green, *Body, Soul, and Human Life*, 58–59.

94. Paul uses *soma*, "body," to refer to our bodies about seventy times in his letters. A good concordance should direct you to some examples.

The Bible I: God and Humans

translation of *soma* would be embodiment: "In this sense soma is a relational concept. It denotes the person embodied in a particular environment. It is the means by which the person relates to that environment, and vice versa. It is the means of living in, of experiencing the environment. But soma as embodiment means more than my physical body: it is the embodied 'me,' the means by which 'I' and the world can act upon each other."[95] Dunn cites a corollary of our bodily nature, namely, our relatedness. Our body is the only way we can *be* in the world and the only means by which we can relate to other humans, to other creatures, to God's world, and to God. This brings up our *embeddedness* and our *dependency*, which we will examine shortly.

Contingency

Our human existence is understood by Scripture as *contingent*. This means that our existence both as a species and as individuals is a "possible occurrence" that was "not certain to happen." Things could have been otherwise. Our existence is "dependent on, associated with, or conditioned by something else."[96] In short, *we are not necessary*.

The psalmists understood this well: "You sweep people away in the sleep of death; they are like the new grass of the morning—though in the morning it springs up new, but by evening it is dry and withered" (Ps 90:5–6). Or as the prophet Isaiah put it: "The grass withers and the flowers fall, because the breath of the Lord blows on them. Surely the people are like grass. The grass withers and the flowers fall, but the word of our God endures forever" (Isa 40:7–8).[97]

In fact, the Bible sees all of creation as contingent. "We live in a contingent universe; that is, it is dependent on something outside of itself for its beginning and its continuing existence."[98] That "something outside" upon which we and all of creation depend is *God*, the only noncontingent, necessary being in existence.[99] This idea becomes concrete when we recognize, as Dr. Nancey Murphy notes in the foreword

---

95. Dunn, *Theology of Paul*, 56.
96. Gove, *Webster's Dictionary*, 493.
97. See also Pss 103:13–16; 104:27–30; Isa 51:12–13; Matt 6:25–34; Luke 12:13–21, 22–34; Jas 1:10–11; 1 Pet 1:24.
98. Achtemeier, *Nature, God, and Pulpit*, 95.
99. Gen 21:33; Pss 90:2; 102:24–28; Mal 3:6; Rom 11:33–36; 1 Tim 1:17; Heb 13:8.

to this book, that if there were slight differences in the forces that hold atoms together, the universe would be very different and life would not be possible. In other words, "the universe appears to be 'fine-tuned' to allow the existence of life."[100] This is a manifestation of God's design, and we are thankful for it, but we recognize that God could have designed it otherwise or could have elected not to create anything at all. God freely chose to create and to design it as he did such that life, including human life, could live and flourish. Furthermore, the continued existence of creation depends on God. The Apostle Paul writes, "He is before all things, and in him all things hold together" (Col 1:17). All things depend on God in Jesus Christ both for their origin and for their continued existence.

Understanding the contingency of creation helps us understand God's love. If God was not obliged or compelled to create (or to sustain) the universe, then why did he do it? Because God is love (1 John 4:8, 16). In other words, creation's existence is "an absolute gift."[101] Our human existence, yours and mine, is an absolute gift, too—a product of God's superabundant love. In love, God wills our continued existence through willing the existence of the whole of his creation of which we are a part.

God is necessary; creation is not. Insofar as we can understand God—or rather, insofar as he has revealed himself to us—God is eternal, necessary, certain, and unchanging in terms of his existence and character. He simply *is* (Exod 3:14; John 8:58). All else is mortal, contingent, uncertain, and changing. Neither creation nor we humans are, by virtue of our nature or any internal attribute of our own, eternal, permanent, or certain. We are all dependent, vulnerable, fragile, changing. When we discuss the so-called new ecology in chapter 6, we will find that ecological science has come to affirm this vital theological understanding.

We may receive eternal life in Jesus Christ (John 3:16), but this life is a gift in the absolute sense (Rom 6:23). There is nothing that we have done or can do, nor is there any inherent attribute belonging to us that inevitably and with certainty entitles us to receive eternal life. It is freely given to us by God; we may receive the new birth, but it comes entirely from and is wholly dependent upon God (1 Pet 1:23). All of us

---

100. Murphy and Ellis, *On the Moral Nature*, 51–52.
101. Schwöbel, "God, Creation and the Christian Community," 164.

humans and all creation owe our moment-by-moment existence and life eternally to God.[102]

This idea of our contingency is a difficult pill for us modern humans to swallow. In our relentless pursuit of deity, we crave certainty, security, and safety. We prefer to deny our contingency. We want to be necessary; we want to be eternal; we want assurance of our permanent existence; and we want a guaranteed escape from the prospect of non-being. But alas, it is not so. Like the grass of the field, we may be here today and gone tomorrow (Ps 103:15–16), vanished without a trace. God alone stands between us and our annihilation. God alone stands between the universe and its annihilation. All depends upon God; all comes from God; all is sustained by God; and all returns to God (Rom 11:36).

I am not suggesting that humans cannot make choices that have eternal consequences, nor that they cannot take some credit for virtue, creativity, and thought. But even our choices are always dependent on God, and as we take credit for our accomplishments, we must always remember whence our powers to achieve those accomplishments come. They are *gifts* of God (Jas 1:16–18). A truly biblical view of self and of creation is that everything constitutes a contingent gift of the Creator. A true grasp of this faith leads directly to the ecological virtues of *humility* and *self-restraint*.[103] I will return to these virtues in chapter 7.

Embeddedness

By embedded I mean that *we are a part of creation*. We are integral to the eco-physical order of the world.[104] We cannot exist otherwise. We belong here. We are made for creation and creation is made for us. We are "natural." This is our home.[105]

Chris Wright notes that "the opening chapters of the Bible do not immediately emphasize human uniqueness. On the contrary, it seems that at point after point the Bible tells us that we have more in common

102. Moreland and Rae, *Body and Soul*, 22.
103. Bouma-Prediger, *For the Beauty of the Earth*, 145–48.
104. Kelsey, *Eccentric Existence*, 213.
105. Jung, *We Are Home*, 69–70. Numerous passages support this idea: Gen 1–2; Pss 8; 19; 74; 104; Isa 40:12–31; 45:9–13; 48:12–13; Jer 27:5; 32:17; Prov 3:19–20; 8:22–31.

with the rest of the animate creation than in distinction from it."[106] Echoing Lawson Stone cited previously, Wright lists these points: (1) The blessing to be fruitful and fill the earth is a blessing shared with the fish of the sea and the birds of the air.[107] (2) Humans are created on the sixth day with "livestock, wild animals, and creepy-crawlies." (3) We are formed from the ground just like all the animals (Gen 2:19). (4) We share the "breath of life" and our fundamental material being with the other creatures.[108] (5) A human is "a living being," an existence shared with other creatures.[109] These truths of Scripture are completely consonant with the findings of modern science that we are material beings. In the theological language of Colin Gunton, "We are ontologically continuous with the remainder of the created order."[110]

The Apostle Paul affirms our solidarity with creation. In Romans 8:19–23, Paul writes that in our sufferings, we groan with creation as we await together with it our liberation from bondage to decay and our "adoption, the redemption of our bodies." Here Paul sees the eschatological unity of humans with creation as we, with all creatures, await the coming of Christ and the renewal of all things (Matt 19:28). Similarly, we see Paul's understanding of our solidarity with creation in terms of redemption: "For God was pleased to have all his fullness to dwell in him [that is, Christ], and through him to reconcile to himself all things, whether things on earth or things in heaven, by making peace through his blood, shed on the cross" (Col 1:19–20). We are one of the "all things" that are encompassed by the redemption found in the blood of Christ. In creation, in ongoing existence, and in redemption, we humans are part and parcel of this earthly creation.

---

106. Wright, *Old Testament Ethics*, 117.

107. It is likely that the blessing in Gen 1:22 is not confined to the fish and the birds. It is a synecdoche meaning that God blesses all his creatures, fish, birds, land animals, plants, and all living things to thrive, multiply, and fill the earth.

108. Gen 1:30; 6:17; 7:15, 22; Ps 104:29–30.

109. Gen 1:20, 24, 28; 6:19; Wright, *Old Testament Ethics*, 117–18.

110. Gunton, *Triune Creator*, 211.

Limitedness

"The fear of the Lord is the beginning of wisdom" (Prov 9:10).[111] This passage points to the biblical understanding that human wisdom begins in the presence of God with "the acceptance of [our] limitations of human knowledge and power."[112] Biblically, we humans are circumscribed beings. "Human knowledge remains limited due to the restrictions set by our bodiliness."[113] In other words, Scripture teaches that wisdom begins with recognition of our limitations—in all respects. We are *bounded beings* in terms of both our capacity to know the world and to act within it.

However much American culture wants to see personal freedom in terms of limitless options, and notwithstanding the modern quest for omnipotence, biblically speaking, we are and always will be *limited beings*. Theologically, nothing else is possible, and as we will see, ecological science affirms this. It is vital that we grasp it and take it to heart. In concrete terms, it includes our limited capacity to know God, to know one another, to know other creatures, to know ecosystems, even to know ourselves. Similarly, we cannot do anything we want, to anything we want, in whatever way we want, whenever we want. We can only act within the limits of eco-physical reality. Our limitedness includes the possibility of error as well as our mortality.[114] The limits and fallibility of our reason is an aspect of our creaturehood. Human knowledge is always partial, contingent, and limited (1 Cor 13:9, 12). Biblical wisdom grasps the profound limits of our human capacities to know and to act within God's creation. As the psalmist affirms, to all human perfection there is a limit, but God's commands are boundless (Ps 119:96).[115]

Human Relationality

As we noted in the last chapter, science confirms that we humans are inherently relational beings and that our relationality involves not only our social world but also the ecological world that surrounds and

---

111. See also Job 28:28; Prov 9:10.
112 Conradie, *Ecological Christian Anthropology*, 147.
113. Ibid.
114. Westermann, *Elements of Old Testament Theology*, 96.
115. See also Num 34:1–10; Job 14:5; Ps 74:17; Hos 4:2.

contains us. Christian theology also confirms this. Our very being is bound up in our relationships. We exist in relationships with other animals and with the ecosystems in which we live. Chris Wright describes a triangle involving God, Israel, and the land (Figure 1). He comments: "The land in which Israel lived in Old Testament times was a reality of major importance in their relationship with God. It was not merely the place where they happened to live. Nor was it merely an economic asset, essential for agricultural viability. Still less was it considered or handled merely as real estate, property that could be bought and sold commercially. The land, for Israel, was a matter of central theological and ethical importance."[116] Wright's insight is helpful because it teaches us at least three things: (1) we humans are related to both God and to the land; (2) God has a direct relationship with the land apart from us humans; and (3) the land was integral to God's relationship with Israel and was bound up in the moral life of the people.[117] The social, economic, political, and religious life of the people impacted the land.[118] Gerhard von Rad writes, "While present-day man lives his life very much isolated from the world, and is determined by the feeling of otherness and foreignness to it, Hebrew man felt it to be much more personally related to himself."[119] In our modern worldview, the physical creation is theologically and ethically marginalized and secularized. It is largely viewed as a commodity to be traded based on the calculus of economic utility. By contrast Wright and von Rad remind us that biblically, the land is much more than a mere commodity. It is God's land, and our own being is bound up with it. We exist in relationship with the land, and this relationship has moral and spiritual meaning for our lives and is integral to our relationship with God.

---

116. Wright, *Old Testament Ethics*, 76.
117. Ibid., 96.
118. Lev 18:24–28; Deut 5:32–33; Prov 2:21–22; Jer 12:4: Hos 4:1–3.
119. von Rad, *Old Testament Theology*, 428.

The Bible I: God and Humans

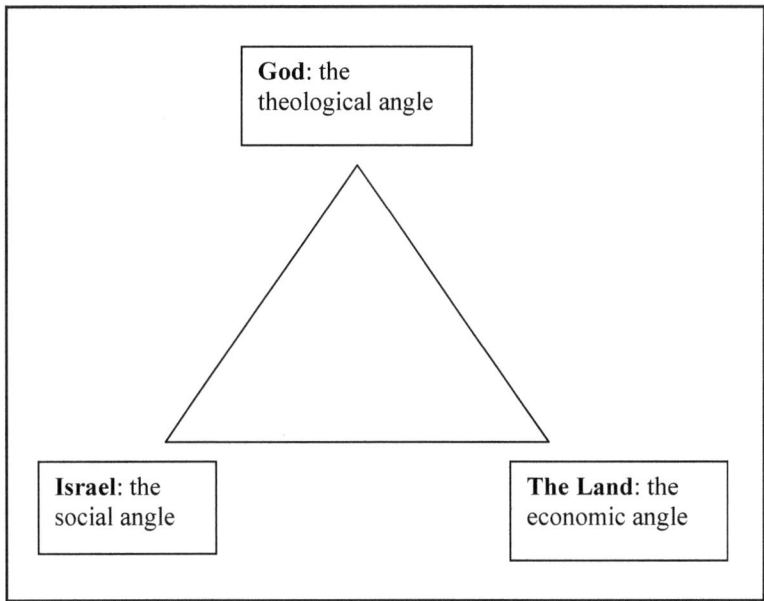

Figure 1: Chris Wright's triangular view of the relationships that inform the theology and ethics of ancient Israel.[120]

Old Testament scholar Daniel Block broadens the categories of Wright's triangle, replacing the people with "all living things" and the land with "the earth" (Figure 2). As he puts it: "The primary cosmic relationships involve God, the world and all life on earth ... From the beginning God intended a trialogical symbiotic relationship in which each of the members responds to the other two in a dynamic covenantal relationship."[121] In this way, he presents a more holistic picture. God is in relationship with all creatures (including us) and with the earth. We are integrated with "all living things." But we also, of course, play a key role as God's governors of his earth. Block goes on to note our central role as God's "deputies and representatives" commissioned by God to care for his world,[122] but the point is that our relationality, including our relationship with God, is bound up in our eco-physical relationships—with the earth as well as with God and other humans.

---

120. Wright, *Old Testament Ethics*, 19 (used by permission of InterVarsity Press).
121. Block, "To Serve and to Keep," 123–24.
122. Ibid., 128.

Dusty Earthlings

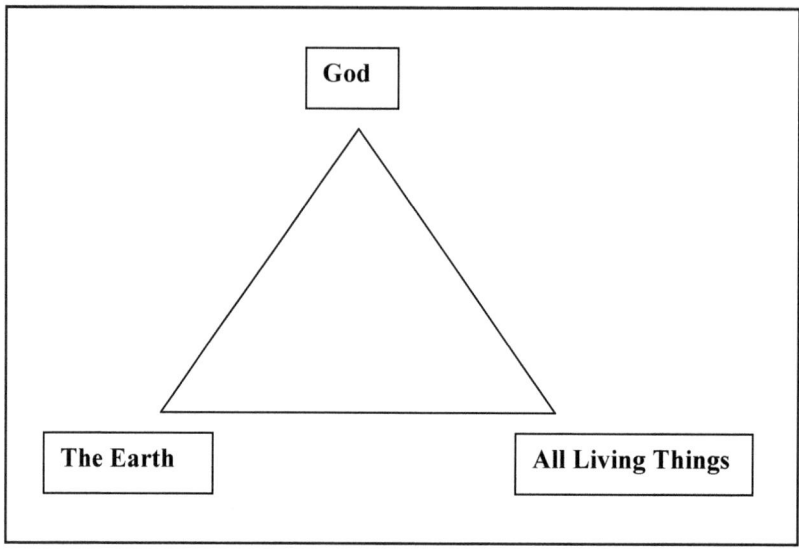

Figure 2. Daniel Block's broadening of Wright's triangular view.[123]

Finally, Christian biologist Calvin DeWitt suggests a triangular set of relationships involving God, other creatures, and humans. He says that "incorporating this third dimension brings greater robustness to the human creatures-other creatures, or humankind-other kinds, relationship. It co-orders both human and other creatures to their Creator."[124] New Testament scholar Richard Bauckham suggests a "four-sided figure, crossed by lines that link opposite corners." At the four corners he places God, humans, other living creatures, and the inanimate creation.[125]

The point of all this is to help us recognize three things about our relationality: (1) God (not us) is supreme in all relationships; (2) God has a direct relationship with creation, the land, and his creatures (apart from us); and (3) we humans are directly related to other creatures, to the land, and to all of creation. In other words, we should remember that our relationality does not consist simply of our relationships with God and with other humans. It involves other creatures and the earth as well. These ecological relationships are vital to our being and our spirituality.

123. Ibid., 123 (used by permission of InterVarsity Press).
124. DeWitt, "Behemoth and Batrachians," 297.
125. Bauckham, *Bible and Ecology*, 146. See also Fretheim, *God and the World*, 13–22.

The Bible I: God and Humans

As we will see in chapter 6, these relationships involve our integration into the matrix of the ecosystem and the exchange of energy, matter, and information that constitutes the dynamics of biophysical life on earth.

The Book of Job: God Puts Us in Our Place

The book of Job is a profound and complex colloquy on God, contingent earthly life, and the existential problems of evil and suffering. Most of the book consists of discussions between Job and his friends about "the moral order of the world,"[126] but at the end of the book we find its climax—God's speech to Job out of the storm (chs. 38–41). In this speech God poses a series of questions to Job about his creation. "Have you journeyed to the springs of the sea or walked in the recesses of the deep? Have the gates of death been shown to you? Have you seen the gates of the shadow of death? Have you comprehended the vast expanses of the earth? Tell me, if you know all this" (38:16–18). "Does the hawk take flight by your wisdom and spread its wings toward the south? Does the eagle soar at your command and build his nest on high?" (39:26–27). The answer to all these questions, of course, is no. Neither Job nor any human could know or do these things.[127] Only God could. God seems to be emphasizing the vast chasm of knowledge and power that separates humanity from him. Job (and we) are creatures who are very different from God.

Surveying "the diversity of life, in which creatures are 'valued' by their creator for their own sake, not for their usefulness to humans,"[128] God shows Job that there is a much larger creation, far beyond the bounds of Job's knowledge or experience, with which God is involved. Job and we are challenged to expand our spiritual and moral vision beyond the bounds of our human personal-social-economic world (which in reality is quite small) to include the entire creation in which

---

126. Bauckham, *Living with other Creatures*, 8.

127. With modern science and technology, we can answer affirmatively to some of God's questions to Job, and we are capable of doing some of the things with which God challenges him. But this does not nullify the meaning of the passage. In Job's time and context, he could not do these things. Thus the point is made: we are creatures and not God. Today, perhaps God might ask us if we have passed through a black hole or journeyed to the center of the sun—things that we cannot and almost certainly never will be able to do.

128. Wilkinson, "Meaning of 'Value' in Biodiversity," 13.

God has placed many creatures that are related to and valuable to God but of no use to humans.[129] We are challenged to broaden our worldview so that we understand ourselves as part of a larger whole over which God is Lord. Similarly, the mention of powerful creatures such as Behemoth (Job 40:15–24) and Leviathan (Job 41:1–11)[130] shows Job (and us) that our power is limited.[131] Although we humans are made in his image, we are *not* God, who made these creatures and can do with them as he wills. We did not make them and cannot do with them as we will. Again, the message is that we are humans *within* (not outside of) God's creation.

Job 38–41 represents perhaps the clearest instance where Scripture puts us humans in our place *within* creation. In effect, God says to Job (and to us): "I am God; you are not. You are a creature among creatures, and there is an infinite difference between me and you." This constitutes a direct challenge to modern humanity's zealous quest for knowledge and ultimate power. Bible scholar David Horrell writes, "God appears to stress, over and over again, all the diverse aspects of the created world which Job does not understand, does not control, and has not witnessed. Moreover, these manifold creatures and processes of the world exist and are sustained by God without any reference to human presence—Job's or anyone else's."[132] Or as Bill McKibben puts it, "We are put . . . in our proper place, not at the center of all affairs but as one part of the larger whole."[133] Richard Bauckham suggests that the effect of God's "barrage of cosmic questions on Job (and the readers of the book) must surely be *cosmic humility*. Before the immensity and mystery of the creation we know ourselves to be creatures whose own place in the world is *limited*."[134] We are not "the unique reference

---

129. Newsom, "Moral Sense of Nature," 23, 25.

130. In Hebrew, Behemoth means "beast of excellence." It may refer to the hippopotamus or the elephant. Leviathan may refer to the crocodile that in ancient times lived along the Nile in Egypt and the Jordan River in Palestine. Expansion of human populations and habitat destruction have eliminated these animals from these areas today. According to Richard Bauckham, most scholars today believe that these refer to monsters of ancient myth (*Bible and Ecology*, 55).

131. Newsom, "Moral Sense of Nature," 24.

132. Horrell, *Bible and the Environment*, 58.

133. McKibben, "Spitting in the Face of God," 24.

134. Bauckham, *Bible and Ecology*, 44; italics mine.

point for all God's purposes in his creation."[135] Herein lies eco-physical wisdom and virtue.

In order to illustrate this point, I present a few diagrams. As I discussed in chapter 1, our understanding of ourselves in relation to the rest of creation (the natural world) forms part of our worldview. The worldview we have been given from our culture sees humanity not as part of God's creation but outside of it, above it, transcending earthly ecological life.[136] Figure 3 illustrates this.

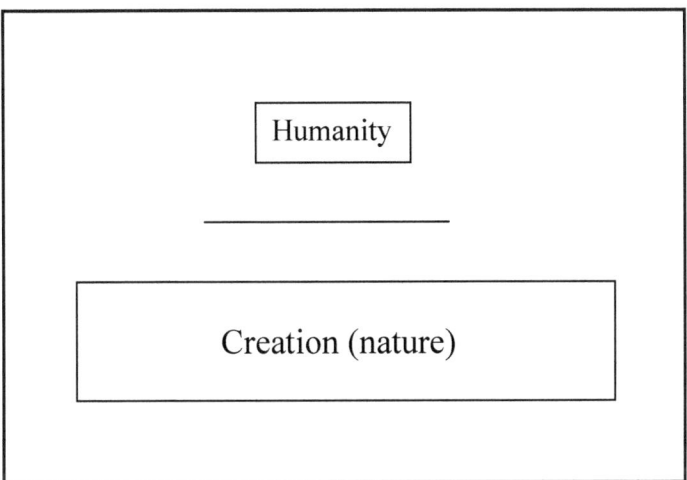

Figure 3. Modern secular worldview that sees humans as separate from and above creation.

As you can see in the figure, we humans are conceived of as above and outside of creation. We can use it, manage it, work on it, control it, study it, contemplate it, and enjoy it, but we perceive it as separate from ourselves.[137] Another aspect of this worldview is that God is absent. This, of course, is consistent with our secular culture, which has excluded God and replaced him (or tried to replace him) with humanity—us.

We Christians have adopted this view and also see ourselves as outside of and above the natural world, occupying an intermediate position between God and creation. We understand that we are different from God, but we also see ourselves as different from creation. This is

---

135. Bauckham, *Living with Other Creatures*, 8.

136. Francis Schaeffer discusses similar ideas. See *Pollution*, 49–50.

137. For a secular, postmodern critique of our perceptual separation from nature, see Cronon, *Uncommon Ground*.

Dusty Earthlings

illustrated in Figure 4. Here the relationship is the same as in Figure 3 except that we include God. The horizontal lines represent our perceived ontological differences between ourselves and creation as well as between ourselves and God. Furthermore, the vertical line shows that we see ourselves as mediators of God's relationship with creation. God relates with creation through us and vice versa.

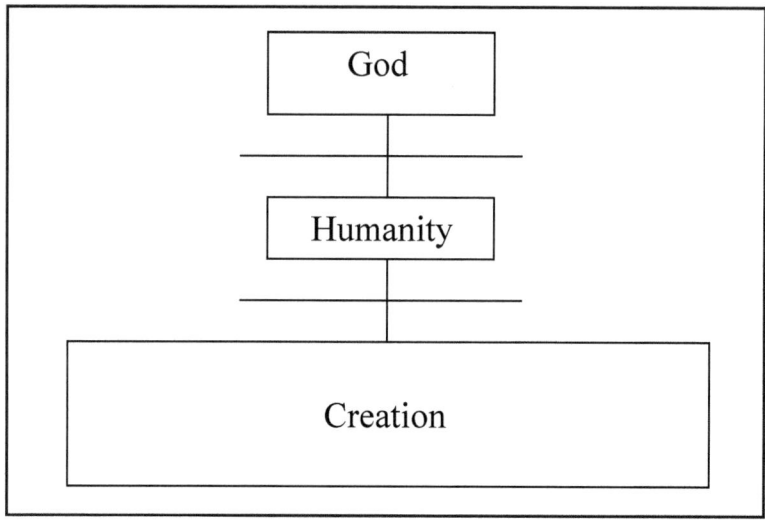

Figure 4. Commonly held Christian worldview that sees humans occupying an intermediate position between God and creation.

I suggest a worldview that is both more biblical and sees humanity as part of creation. This is depicted in Figure 5. In this view, we humans are in the same category of being as the rest of creation. The horizontal line again represents an absolute ontological difference between God and creation (the God-creation dualism I mentioned earlier), and we humans are located below that line, within creation.

The Bible I: God and Humans

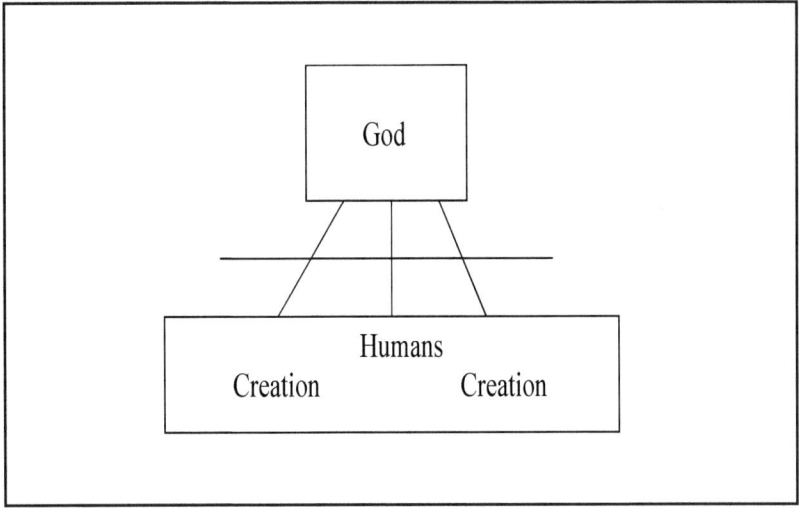

Figure 5: Worldview that sees humans as part of creation, all under God.

This figure also shows that God has a direct relationship both with us humans and with all of his creation. When we see a fox, a tree, a river, or a mountain, we recognize (as Job was persuaded to do) that these things are directly related to God apart from us. But you can see that I have placed humans in the middle and at a higher level than the rest of creation. This represents the higher value and position that humans occupy within the order of creation (Ps 8:5). This includes our role as caretakers (Gen 1:26–28) and our uniqueness as the ethical species (chapter 3). We are unique and yet we are of a piece with creation.

A Holistic View of Creation

"In the beginning God created the heavens and the earth."[138] That is to say, God created the *whole* of creation. As Old Testament scholar Claus Westermann says, "The intention of the first chapter of the Bible is . . . to understand *the world as a whole in view of the creator*."[139] In other words, we must understand the primordial status of creation vis-à-vis God as holistic—the entire creation in relation to God. Humanity does not stand alone but is a part of this whole (see Figure 5.)

---

138. See also Neh 9:6; Ps 146:6; Isa 45:18; 66:2; Jer 10:12; John 1:3; Acts 17:24.
139. Westermann, *Elements*, 87.

Westermann goes on to point out that a major purpose of the Genesis narrative is to help us humans "understand the world as a whole. The world as a whole is not accessible to [our] senses; an individual person could always only perceive a tiny section. The whole of the world could only be comprehended from the perspective of its origin."[140] As humans we cannot see the world all at once as God can. Whether working, socializing, enjoying nature, or engaging in scientific study, each of us can see only a tiny fraction of all that is. But in the creation narrative, God presents a broader view. Some have suggested that the famous photograph of the earth taken by Apollo 8 astronauts from moon orbit in 1968, showing the earth as a beautiful, blue-and-white sphere suspended in space, was a momentous revelation of the earth as a unit and ourselves as part of it. For the first time in history, some people declared, humans were confronted with their world as a whole. That may have been true for some, but for those who read their Bibles, long before that picture was made, the Genesis narrative shows us this holistic view.

Westermann also notes that this holistic view of creation has been overlooked by theologians. He points out that our anthropology has tended to be an abstract one—of humans in their "bare existence" in relation to God alone, as "individuals abstracted from their vital relationships."[141] This is an understatement. At least in the modern era, theology has viewed humans as almost exclusively personal, as isolated individuals existing in relationship to God and sometimes to other humans but almost totally removed from eco-physical relationships. Not only in Genesis 1:1, but later at the end of the flood narrative (Gen 6–9), we see this holistic idea again. God makes a covenant with humans *and* all creatures: "I now establish my covenant with you and with your descendants after you and with every living creature that was with you—the birds, the livestock and all the wild animals, all those that came out of the ark with you—every living creature on earth" (Gen 9:9–10). This covenant of God with all creatures shows that the Bible embraces living creation as a whole and humans as a part of that.[142] James Nash notes that this covenant and the divinely directed act of preserving all the animals affirm the idea that humans are a part of a larger community of

---

140. Ibid., 90.
141. Ibid., 95. See also Fretheim, *God and the World*, ix–xi.
142. Bauckham, *God and the Crisis of Freedom*, 176.

living creatures.[143] Loren Wilkinson sums up: "Biblical theism portrays humanity in a web of 'horizontal' relationships to other creatures (a portrayal deeply resonant with current ecological understanding)."[144] This holistic idea of creation has a number of implications for how we think and live in the world.

First, a holistic worldview means that although we may be members of a particular family, local community, race, ethnic group, or nation, we see ourselves primarily as God sees us—as members of the worldwide human family. As C. S. Lewis aptly puts it in his *Chronicles of Narnia* stories, we are all sons of Adam and daughters of Eve. But we are not only embedded within the human family, we are embedded within the larger created order—the eco-physical "family" of creation.

Second, it means there is no division between spiritual and material. "Through him [Christ] all things were made; without him nothing was made that has been made" (John 1:3).[145] Thus, through and in Christ the whole of existence came into being, in both its spiritual and physical aspects—angels and animals, prayers and plants, communion and cougars, baptisms and butterflies. This means that any worldview that splits the world into spiritual and material realms misunderstands the integrated wholeness of created existence. All that is not God is "one created reality, visible and invisible."[146] To exist spiritually is to exist physically, and to exist physically is to exist ecologically. All aspects of created being are integrated aspects of holistic being.

Third, it means that ecologically we are part of a larger eco-physical whole, integrated members of the worldwide ecosphere. Its biotic and abiotic components, its systems, cycles, and webs, were created by God. Thus we should seek to live out our membership within this larger whole and preserve it because we are part of it.

---

143. Nash, *Loving Nature*, 100–101.
144. Wilkinson, "Meaning of 'Value' in Biodiversity," 13.
145. See also Col 1:16.
146. Harris, "Environmental Concern," 13.

Jesus Christ: God in Human Form, the Archetype of Embodied Humanness

The strongest biblical affirmation of created physicality rooted in earth is Christ himself. In the *incarnation* ("enfleshment") of God, the physical crucifixion, and his physical resurrection, God himself certifies the value and significance of the physical creation and of humans as a part of it. The incarnate Christ was a dusty earthling, a part of the earth's crust—just like us.[147] As theologian David Kelsey notes, Jesus of Nazareth was a "living, genetically human body"[148] who became hungry, thirsty, tired; who rejoiced, grieved, and became angry; who suffered fear, anger, humiliation, heartbreak, pain, and death. As theologian Arthur Peacocke puts it, "the man Jesus existed in the world as we do; he was made of carbon, nitrogen, oxygen, and so on, as we are . . . His DNA was patterned on the same genetic code as is ours, and as is that of all other living creatures from microbes to animals."[149] In his life and ministry, Jesus was very physical. Several passages refer to bodily activities—eating, walking, drinking, and so on.[150] His healing of physical problems affirms that the physical healing of human bodies is within God's redemptive plan. In healing the eyes or the legs or the brains of people, he was healing persons. Since Jesus was physical, he was also ecological. As fully human, Jesus was an eco-physical being.

Eastern Orthodox theologian Paulos Mar Gregarios observes that we Western Christians have tended to spiritualize Christ, to see him only as a spiritual being. He affirms Christ's physicality, even beyond the resurrection:

> Jesus Christ is not an abstract or "purely spiritual" entity. He is incarnate. He took a material body, becoming part of the created order while remaining unchanged as one of the three persons of the Trinity who is Creator. He is one of us. He is fully consubstantial with us . . . Christ the Incarnate One assumed flesh—organic, human flesh; he was nurtured by air and water, vegetables and meat, like the rest of us. He took matter into himself, so matter is not alien to him now. His body is a material body—transformed, of course, but transformed matter. Thus

---

147. John 1:14; Heb 2:14, 17. McClendon, *Ethics*, 89.
148. Kelsey, *Eccentric Existence*, 1013.
149. Peacocke, *Creation*, 244–45.
150. See for example Matt 11:19; Mark 1:16; 14:22–25; Luke 24:15, 43; John 10:23.

> he shares his being with the whole created order: animals and birds, snakes and worms, flowers and bees. All parts of creation are now reconciled to Christ.[151]

Understanding the dustiness of Jesus Christ—his incarnate physicality—is difficult for us. The Docetism within modern Western Christianity is deeply embedded in our minds and souls.[152] Perhaps it requires a Gregarios, someone from outside our culture and tradition, to call our attention to it. It is as hard for us to understand Jesus as an eco-physical being as it is for us to understand our redemption as being embedded within the creation. These are difficult concepts for us, habituated as we are to a spiritualized conception of Jesus as "our personal Savior." Notwithstanding the importance of Jesus' saving grace for each one of us, we must also understand that his incarnation, death, and resurrection have a much broader impact. Just as through Christ all things were made (John 1:3), so also through Christ all things are redeemed to the Father (Col 1:19–20). Jesus Christ, the incarnate God, affirms eco-physical nature.

---

151. Gregarios, "New Testament Foundations," 88–89.

152. Docetism was an early heresy that taught that Christ did not have a body but only seemed to have one. This view of Christ has cropped up in various forms several times in church history.

5

The Bible II: The Image of God, the Dominion Mandate, and Other Issues

HAVING DISCUSSED BIBLICAL IDEAS about who God is and who we humans are in relation to God and to the rest of creation, in this chapter I want to address some additional biblical topics that are relevant to these issues and to the argument of the book. These include the image of God, the dominion mandate, ecological sin, redemption, and theocentrism.

Humanity Made in the Image of God[1]

"Then God said, 'Let us make humankind in our image, according to our likeness'.... So God created humankind in his image, in the image of God he created them; male and female he created them" (Gen 1:26–27, NRSV).[2] Old Testament scholar Gerhard von Rad claimed that the image of God in humans is actually a minor issue in the Hebrew Bible: "The central point of [Old Testament] anthropology is that man is dust and ashes before God and that he cannot stand before His holiness. Thus the witness to man's divine likeness plays no predominant role in the [Old Testament]. It stands as it were on the margin of the whole complex."[3] So according to von Rad, the key feature of our humanity in the Bible is our dustiness, our solidarity with the physical creation, and our ontological and moral difference from God. Furthermore, we should note that in the New Testament the image of God focuses on

---

1. For more detailed discussions of the image of God in humans in relation to ecology, see Middleton, *The Liberating Image*, and Hall, *Imaging God*.

2. See also Gen 9:6; Ps 8:5–6; 1 Cor 11:7.

3. von Rad, "*eikon*," 2:390.

Jesus Christ, not on humanity in general. It is Christ incarnate who *is* the true image of God,[4] and humanity is subordinated to that.[5] But despite these biblical realities, the meaning of the image of God (in Latin *imago Dei* or simply the *imago*) in us humans has been a much discussed issue for Christian anthropology. Since this issue is key to our understanding of who and what we are as humans, which is a central topic of this book, I need to discuss it. How can we be dust, inherently physical beings, yet at the same time be *in the image of God*, when God, so far as we know, is a nonphysical being—or perhaps better a "supra-physical" or "trans-physical" being—who is radically different from us? I cannot, of course, offer final answers to these questions or to the question of what exactly is the image of God in humans. But I will offer some ideas that, I hope, will help us understand how, within the limits of our eco-physical nature and context, we can exist in the image of God. Before I proceed, I would like to offer four caveats.

(1) Being Made in the Image of God Must Be Understood in Light of Our Physical Nature

As I stated in chapter 4, my approach to Scripture presupposes that we are eco-physical beings. This means that however we interpret the image of God in humans, we should understand it in a physical way. Just as we bring to Scripture the understanding that the earth is a sphere that rotates on its axis and orbits the sun, we also bring our eco-physicality to it as a fact of our existence. There is ample scriptural support for this. Again, if you are a body-soul dualist or you ascribe to some other anthropology, I am not suggesting that you abandon that. All I am asking is that you take the fact of our eco-physical existence into account as we seek to interpret Scripture and its "image of God" language. If we keep this in mind, we will, I believe, move closer to a "right" interpretation of these passages, an understanding that is corroborated by what science tells us about ourselves.

---

4. 2 Cor 4:4; Col 1:15; Heb 1:3.
5. Shults, *Reforming Theological Anthropology*, 239.

(2) The Image of God Should Be Interpreted Holistically

I suggest that the *imago* refers to the human being as a whole—all of me and all of you—body, brain, mind, heart, arms, legs, chest, stomach, consciousness, personality, relationships, history, faith, hopes, and so on. Historically, the image of God has sometimes been associated with certain parts of our being—our rationality, our moral capacity, or the soul. Augustine, for example, believed that the image of God was found in the soul, specifically the capacity of the soul to reason. "It is in the soul of man, that is, in his rational or intellectual soul, that we must find that image of the Creator."[6] This view is no longer tenable. The image of God must apply to all of our being, bodies included.[7] As theologian Colin Gunton puts it, "Our *particular* embodiment . . . is inseparable from our being in the image of God."[8] The image cannot be spiritualized or intellectualized. The whole person is created in the image of God.[9]

(3) The Image of God Is the Theological Basis for Our Uniqueness within Creation

Within the framework of Genesis 1, the image of God is what designates us humans as unique among God's creatures. Whereas they are created "according to their kinds" (Gen 1:21, 24, 25), we are created in the image of God.[10] Thus, while the *imago* must be interpreted holistically, it must also signify that which is unique about our humanity. We find Scripture (and science) to be consistent on this point. We are

---

6. Augustine, *On the Trinity*, xiv, 4, 6, quoted in Niebuhr, *Nature and Destiny of Man*, 154–55; see also Moltmann, *God in Creation*, 236–38.

7. I am not suggesting here that our bodies actually resemble God in some way. Scripture contains some theophanies in which God is represented in some physical-like form, but I don't know what God "looks" like (and I don't think anyone does). What I am suggesting is that our interpretation of the image of God must be taken in such a way that it applies to all aspects of our being—the whole human person including our bodies. I am also not suggesting that persons who lack certain body parts, such as amputees, or who are disabled in some way do not bear the image of God. Again, the idea here is not exclusion but inclusive holism.

8. Gunton, *Triune Creator*, 205.

9. von Rad, *Genesis*, 58.

10. Berry and Jeeves, "Nature of Human Nature," 25; Goldingay, *Old Testament Theology*, 2:518.

The Bible II: The Image of God, the Dominion Mandate, and Other Issues

eco-physical beings (like other animals), but our being in the image of God makes us unique.

(4) The Image of God Does Not Make Us the Central Focus of Creation

Although we are a unique species within creation, this does not mean that we are the center of the universe nor that all else exists simply for our sake. Unfortunately, the *imago* has occasionally been used to support this view. Creation exists for the sake of God, not us. This idea is called *theocentrism*, "God-centeredness," which I will discuss later in this chapter. It is a difficult concept to grasp, let alone live. But grasp it we must, if we hope to correctly interpret the image of God.

Three Interpretations of the Image of God

Having offered these four caveats, I now want to review some traditional interpretations of the *imago*. They generally fall into three categories: *substantive*, *relational*, and *functional*.[11]

A *Substantive* Interpretation

This says that the *imago* refers to some thing or substance intrinsic to humans. As we just saw in Augustine's interpretation, the *imago* can be interpreted to mean the intellect, mind, or soul. These are generally defined as a nonmaterial or "spiritual" substance or entity that is joined with a material body to form a whole human being. In light of my claim that humans are essentially or ontologically physical beings, this interpretation would either be ruled out or could only be viewed holistically as referring to both soul and body, the "total psychosomatic person as the image of God."[12] If we do accept a substantive interpretation, then the *imago* must refer to our physicality in some way—perhaps to some general aspect of our bodies and brains. In chapter 1, I observed that there is an identity, a personality, pattern, or paradigm—an organizational structure of information and processes—that identifies, defines,

---

11. Erickson, *Christian Theology*, 520–29.
12. Peacocke, *Creation*, 284.

and expresses the person as a unique being. I suggested that this could be embodied in DNA sequences, in neural patterns and networks, personal history, and so on. If we viewed this as the image of God, then perhaps we could understand this as a substantive interpretation. There are many complications here, but in this sense, perhaps a substantive interpretation is possible. Again, my purpose here is not to argue about human constitution but to affirm our physicality.

A *Functional* Interpretation

This view sees the image of God as something humans *do*—a job or role that humans fulfill within the created order. A common functional interpretation sees humans as fulfilling the role of overseers or managers, exercising dominion over the earth and its creatures (Gen 1:28).[13] Indeed, in Genesis 1:26–28 the image of God language is connected with the mandate to rule over other creatures. It seems to me that this functional idea is explicit in 1:26 where God says, "Let us make humankind in our image, according to our likeness, and let them have dominion over the fish of the sea . . ." (NRSV), and also in 1:28 where humans are given a command to rule over and subdue the earth. Thus, this "dominion mandate" is tied to the image of God. In other words, the image of God carries with it the role or function of being overseers, managers, or stewards of God's earthly creation. The nature of this role and how it should be carried out is an area of controversy among Christians. Discussion of this would fill another book and then some. I will not discuss it in detail, but I will offer some thoughts in the section on the dominion mandate later in this chapter.

Another functional view of the *imago Dei* sees humans as God's representatives on earth.[14] That is, humans represent God to the rest of creation. Old Testament scholar Gerhard von Rad describes the biblical background of this interpretation: "Just as powerful earthly kings, to indicate their claim to dominion, erect an image of themselves in the provinces of their empire where they do not personally appear, so man is placed upon earth in God's image as God's sovereign emblem. He is really only God's representative, summoned to maintain and enforce

---

13. Erickson, *Christian Theology*, 527.
14. Hart, "Genesis," 319.

God's claim to dominion over the earth. The decisive thing about man's similarity to God, therefore, is his function in the non-human world."[15] Note von Rad's use of the word *function*. Humans function as representatives of God to the rest of creation. Old Testament scholars Walter Brueggemann and J. Richard Middleton both say that this interpretation is now generally accepted among scholars.[16] "As *imago Dei*, then, humanity in Genesis 1 is called to be the representative and intermediary of God's power and blessing on earth."[17] Given that love is intrinsic to God's nature (1 John 4:8, 16), our function then is to represent or embody that love toward other creatures. I will say more about this presently when I discuss dominion in the image of Christ.

Another functional interpretation proposed by theologian H. Paul Santmire is that we humans are to be "wondering onlookers" of creation.[18] Santmire believes this is a vital but forgotten aspect of the image of God for us humans. Scripture says that God looked upon his creation and found it *good*.[19] In fact, after he completed his creative work, God looked on all he had made and found it *very good* (1:31). Like an artist who, after completing his work, steps back, looks at it, smiles, and finds it to be just right, God stepped back from his creation and found joy and satisfaction in what he had made. Likewise, we humans are the uniquely gifted creatures who can, in his image, look upon other creatures and the rest of creation and find appreciation and joy in them. We can wonder at them, ask questions about them, learn about them, and consciously join them in offering praise and thanksgiving to God.

A *Relational* Interpretation

This approach views the image of God as constituted in the relationships we humans have, or are capable of having, with God, with one another, and with creation.[20] The Trinity—Father, Son, and Holy Spirit existing in dynamic relationship within the Godhead—is often cited as the basis of human relationality. God is an inherently relational be-

---

15. von Rad, *Genesis*, 60. In reference to this interpretation, von Rad cites W. Caspari, "Imago Divinia," in *Festchrift fur R. Seeberg* (1929) 208.
16. Brueggemann, *Genesis*, 32; Middleton, *Liberating Image*, 27, 107, 121.
17 Middleton, *Liberating Image*, 121.
18. Santmire, *Brother Earth*, 190.
19. Gen 1:3, 10, 12, 18, 21, 25, 31.
20. Erickson, *Christian Theology*, 523–24.

ing, and so are we. This relationality is, first and foremost, expressed in our unique capacity for relationship with God. To love God is our first commandment (Mark 12:30). As Augustine said, "To praise you is the desire of man, a little piece of your creation. You stir man to take pleasure in praising you, because you have made us for yourself, and our heart is restless until it rests in you."[21]

Theologian Glen Scorgie declares, "Human existence is essentially relational."[22] Colin Gunton agrees. Like the persons of the Trinity, we "subsist in mutually constitutive relations" with other people. We are not primarily individuals and cannot exist as such.[23] We are designed for community with other persons.[24] Although sociality is a trait we have in common with other social animals, the depth, extent, and complexity of our social nature is unique among creatures. We are, in many ways, *the* social species. Scorgie and Gunton point to the important truth that we are ontologically constituted by our social relationships, a truth that is being corroborated more and more by the findings of science.[25] Thus human social relationality can be taken as an essential aspect of the image of God in us.

Our relational understanding of the *imago* is deficient, however, if it does not also include our ecological relationships with other creatures and with the ecosphere. Colin Gunton notes, "There is a further element of being in the image of God which derives from our being created persons, and this is our continuity with the remainder of creation and our necessary embodiment."[26] This is not an afterthought. It is, I suggest, an integral part of our being in the image of God as dusty earthlings. We

---

21. Augustine, *Confessions*, 3.
22. Scorgie, *Little Guide*, 31.
23. Gunton, *The One, the Three and the Many*, 208.
24. Both Scorgie and Gunton are claiming that humans are in essence constituted in community. This is in conflict with modern individualistic understandings of human nature as, for example, in economic theory where humans are viewed atomistically as individuals acting rationally in their own self-interest.
25. See Murphy and Brown, *Neurons*, 72, 77. Science is finding that the "self" or the "person" does not stop at the skin but extends beyond it. A human person is constituted by a complex of ever-changing relationships, both historical and current, with people, creatures, things, places, lands, seas, and so on. From the view of neuropsychology, we *are* our relationships, and we are embedded in our physical, social, and ecological contexts.
26. Gunton, *Triune Creator*, 208.

The Bible II: The Image of God, the Dominion Mandate, and Other Issues

are relationally embedded in the living community of creatures within the ecosphere, and we are constituted by that community.

Relationality, then, should be accepted as a key aspect of the image of God for humanity. We humans possess the unique capacity for active, conscious relationships with God, with other humans, with God's other creatures, and with God's creation.[27] Our principal relationship is, of course, with God through Jesus Christ (John 1:12–13). But in order to live and flourish as whole humans, we require a matrix of *social* and *ecological* relationships. By our very nature, we are obligated to live within this matrix. There is no alternative. Without it human life is not possible—at least not in any meaningful way.

The Image of God Includes All Three Interpretations

I suggest that a holistic understanding of the *imago Dei* recognizes that it is *complex*. Its meaning cannot be confined to any particular aspect of our being, nor can it be understood in terms of any one of the above three interpretive approaches. The *imago* is, or should be, manifested in all that we are and all that we do.[28] This gives us a holistic picture and recognizes our spirituality as well as our embodied, eco-physical nature. I suggest that "substantively," we are created in the image of God in terms of the array of qualities and characteristics that make us unique among creatures—relationality, language, sociality, metacognition, complex culture, conscious moral life, and so on.[29] Functionally, we rule over creation on God's behalf and represent God to his creation.[30] And we live in relationship with God, with other humans, and with all of creation.

The Image of God Is a Gift

In chapter 4, I noted that as biblical Christians, we understand our very existence and the existence of all creation as a contingent gift of God,

---

27. Green, "'Bodies—That Is, Human Lives,'" 172.
28. Jeeves and Brown, *Neuroscience, Psychology, and Religion*, 126.
29. Humans in whom these capacities are damaged or disabled are no less human. They still exist within relationships and reflect the Trinitarian nature of God.
30. Bauckham, "First Steps," 233.

the result of his love. Likewise, the image of God is also a gift—freely bestowed by God and received by us in gratitude and praise. We do not deserve it, and we are not entitled to it. The gift of the image of God is a vital part of a proper theological and ethical understanding of ourselves and our role as humans living in God's earth.

The Image of God Gives Us Power

Being created in the image of God and being given oversight of his creation means we have power. This is evident in the way we humans have changed and continue to change the face of the earth. In a sense, in God's image we mediate God's power to the world.[31] But with that power comes responsibility. "There is a call inherent in every gift. The *imago Dei* is thus inextricably linked to the gift and responsibility (delegated to humanity at creation) of exercising stewardship over the earth."[32] With power comes *risk*. Power tends to corrupt us, as Winston Churchill is reported to have said, and given our limitations in vision and understanding and our sinful nature, there is indeed great risk of error and abuse in our rulership over God's creation. Our ecological track record as a species bears this out. But this divine gift of power is to be used in ways and for purposes that are carefully discerned in Scripture and in the person of Jesus Christ within the framework of created eco-physical existence. God holds us accountable (Luke 12:47–48).

The Image of God Is the Basis of Human Dignity

The image of God confers upon humanity a dignity and value that is unique among God's creatures.[33] However we interpret the *imago Dei*, we should recognize that it affords a fundamental dignity and value to humanity as a whole and to each individual human. The fact that we humans are "in the image of God" means we occupy a special place among the creatures. With this special place comes special value but also, as we have seen, special responsibility for the care and management of God's

---

31. Middleton, *Liberating Image*, 28.
32. Ibid., 204.
33. Ibid., 232.

The Bible II: The Image of God, the Dominion Mandate, and Other Issues

earthly creation. We, you and I, are accountable to God for the condition of his creation and his creatures.

Other Creatures Have Dignity and Value

The fact that God gives to us humans dignity and value does not diminish the dignity and value of other creatures, nor does it mean that the value of other creatures is understood only in relation to humans. Jesus said, "Look at the birds of the air; they do not sow or reap or store away in barns, and yet your heavenly Father feeds them. Are you not much more valuable than they?" (Matt 6:26–27).[34] Jesus' point here is that we humans are of great value in the eyes of God. Thus, we need not worry about our needs because God will care for us. We are, in fact, more valuable than birds and flowers. But his argument is premised on the notion that the birds and flowers are also valuable to God. Jesus' statement would make no sense if the birds and flowers had no value or if they had value only in relation to their usefulness to humans. God cares about the birds and all animals, but he cares about us humans more. God loves and cares for all creatures in a way that is consistent with their natures and with the order of creation (Ps 104:27).

The Goal of the Image of God Is to Be Conformed to the Image of Christ (Rom 8:29)[35]

Theologian F. LeRon Shults declares, "For the New Testament authors . . . the concepts of the image and likeness of God refer directly to Jesus Christ, and our sharing in the glory of the divine image is possible only by our becoming like him and in him."[36] In chapter 1, I said that we Christians have a source for moral guidance in the historical person of Jesus Christ, the incarnation of God in the world, the ultimate dusty earthling, the archetype of humanity as it ought to be (Luke 3:22), or what theologian David Kelsey calls "the grammatically paradigmatic human being."[37] Being made in the image of God means

---

34. See also Matt 10:29; Luke 12:22–31.
35. Gunton, *Triune Creator*, 199.
36. Shults, *Reforming Theological Anthropology*, 239.
37. Kelsey, *Eccentric Existence*, 1008.

that we humans, in contrast to other animals, have the *capacity* to live and behave *like God*, or *in the image of God*, or more precisely, *in the manner and way of Jesus*, God in the flesh. This is what David Kelsey helpfully calls "imaging the image of God."[38] Thus, the *imago Dei* means that Christ *is* the image of God, and we humans have been gifted with the capacity for life and action in that image (John 1:12). Empowered by the Spirit, we can imitate Jesus Christ, and we can be like him in our eco-physical living as managers of God's earthly creation.

In sum, then, the image of God should be interpreted in terms of our eco-physicality and carries complex meaning in terms of substance, function, and relationality. It is a gift and confers upon us dignity, value, power, and responsibility. Although we are of greater value than other creatures, they too have value in the eyes of God. Finally and most importantly, the goal of the image of God is that we live and behave *in the manner and way of Jesus*. In chapter 7, I will attempt to flesh out some of the ethical implications of this.

The Dominion Mandate

"God blessed them [the humans] and said to them, 'Be fruitful and increase in number; fill the earth and subdue it. Rule over the fish of the sea, and the birds of the air and over every living creature that moves on the ground.'" (Gen 1:28). Commonly called the "dominion mandate,"[39] this passage gives to humans power over God's earthly creation. In ancient times when this was written, human power over nature was quite limited compared to today. Even so, Richard Bauckham suggests that the author was simply recognizing the empirical fact that even back then humans were dominant over other creatures and over the ecosystems in which they lived.[40] Indeed, we can see that throughout all of human history, for the most part, we have been in control, and today there can be no doubt that we have largely filled, mastered, subjugated, and conquered God's earthly creation, if that is what was intended by the

38. Ibid., 1011.

39. This refers to the passage in Genesis 1:26–28 in which God commissions humanity to rule over and subdue the earthly creation. It is also sometimes called the cultural mandate.

40. Bauckham, "First Steps," 233.

The Bible II: The Image of God, the Dominion Mandate, and Other Issues

"subdue" language this passage. As theologian Arthur Peacocke says, "Man, whether by intent or not, now finds himself at the helm of his spaceship and his own fate is linked with that of all its systems."[41] Human dominion is an indisputable fact. Today, in this era of massive human impact not only on local ecosystems but on the global ecosphere, the meaning of the dominion mandate has become more important than ever—and also more controversial.

Like the image of God, the dominion mandate has been interpreted in myriad ways, and as with the *imago*, as we seek to discern its meaning, we must keep in mind our tendency to interpret Scripture in service of ourselves rather than of God. This tendency presents a greater peril when we interpret Genesis 1:28. Like children whose parents have gone away and left them the run of the house with only brief instructions as to its management, we are lamentably inclined to interpret God's mandate to manage his household as license to "party it up" and forget our responsibilities.[42] We ought to be on guard against these temptations. As I argued in the last chapter, it's about God, not us. The power given to us by God in the dominion mandate should be interpreted and used in service and honor of God—always. But this is not an easy task, and while I admit that the meaning of Genesis 1:28 is controversial, I still think this goal must be kept foremost in our minds.

Now let's look at this passage. As noted, God here confers upon humanity authority and power over his creation. The verbs for rule, *radah* (vv. 26, 28), and subdue, *kabash* (v. 28), are strong words that denote forceful control and subjugation.[43] On the one hand, some have interpreted this to mean that creation is there for us to do with as we wish. On the other hand, some have sought to mitigate the force of these words by connecting them with Genesis 2:15 where we humans are instructed to care for and watch over the garden (with garden being interpreted as a synecdoche signifying all the earthly creation). I suggest a mediating view. The dominion mandate clearly sanctions human use and modification of the earth for our legitimate purposes. We are biblically justified in farming the land, building cities, developing technology to improve our lives[44] and to control diseases and pests, and using

---

41. Peacocke, *Creation*, 264.
42. Peterson, "In and Of the World?" 240.
43. Goldingay, *Old Testament Theology*, 1:110, 113, 116.
44. In our society of abundant wealth, convenience, and "hyper-development," we

natural resources for our needs. But those who argue that dominion should also be understood in tandem with Genesis 2:15 are also correct. Dominion indeed includes the idea of care, nurture, protection, and preservation of God's nonhuman creation and his creatures. Ultimately, we should take to heart the fact that the gift of dominion is precisely that, a *gift*, and it is not without conditions and constraints. What are those conditions and constraints? I suggest there are at least seven.

(1) Our Power Is Limited and Particular

It is not like God's. This may seem obvious, but we easily forget it, especially in this modern era when our power over nature is so great. Our dominion is circumscribed by our creatureliness—our eco-physical nature, our dependent solidarity with God's earthly creation,[45] and our subjection to the principles, patterns, parameters, and limits of eco-physical existence. We cannot transgress the laws of physics—the law of gravity, for instance. If we attempt to do so, we will pay a price. In the same way, we cannot transgress the "laws" of ecology either, without paying a price.[46] The price may be delayed and be paid in irregular and unexpected ways, and maybe not by the transgressors themselves but by someone or something else, but it will be paid. We cannot do anything we want—or, as Francis Schaeffer puts it, not everything that can be done should be done.[47]

The ecological limits of dominion dovetail with the theological limits. Theologian Elizabeth Achtemeier notes that God wrote the limitations of our existence into the order of creation. The command not to eat of the tree of good and evil (Gen 2:17) expresses the idea of limits for the dominion of humans. "In other words, there is a limit on our existence, a limit beyond which we must not step if we wish to have life abundant. We are creatures and not the Creator, and we are dependent

---

need to ask questions about what "improvement" of our lives means and when we have "improved" them enough.

45. Santmire, *Brother Earth*, 147.

46. As we will see in chapter 6, the idea of fixed laws in ecology is ambiguous and complex. Nevertheless, ecologists have discovered a number of principles, patterns, parameters, and limits. They can tell us a lot about how God's ecosphere works and what we can and cannot do—or at least what the consequences might be.

47. Schaeffer, *Pollution*, 91.

The Bible II: The Image of God, the Dominion Mandate, and Other Issues

on his love for us."[48] Only within the limits of dependence on God and on his creation can we flourish as his people. "That is the other side of the marvelous balance in the Bible's view of human nature, for the limitation prevents every proud attempt to rule nature and history in the place of God."[49]

Indeed, we must not confuse our dominion with that of God's. As noted in chapter 4, there is an infinite difference between us and God. God's dominion surrounds and dominates our own. If we have power to do something, it is only because God has graciously given it to us.

(2) Dominion Is a Gift

Dominion over God's creation is not an entitlement, nor is it merited, nor is it an intrinsic attribute of our nature. It is a *blessing*, a contingent *gift* bestowed by the hand of God (Ps 8:5–9). It is not a possession to which we have a natural right, but a privilege that we gratefully receive and for which we are humbly thankful—a thankfulness that should be borne out in our use of the gift. Furthermore, we are accountable to God for how we use it.[50] Misuse of our power over the earth and its creatures is an affront to God's grace and generosity, a presumption upon his love, and may constitute an attempt to usurp his authority. As with any contingent gift, its abuse may precipitate its loss. A spirituality of humility and gratitude for the gift of dominion recognizes that it may be taken away at any moment. Our power is always derivative, partial, contingent, and limited. This alone is due cause for its grateful, humble, and cautious use.

(3) We Are Insider Stewards[51]

We are not above or outside of nature. We are not transcendent rulers in the way that God is.[52] Yes, we have a measure of control, but not from

---

48. Achtemeier, *Nature, God, and Pulpit*, 58.
49. Ibid., 59.
50. Hart, "Genesis," 324.
51. I am grateful to my friend Dr. Glen Scorgie, who originated this phrase "insider stewards."
52. Bauckham, *God and the Crisis of Freedom*, 128.

above—only from within. We are in and of the earth.[53] As theologian Ernst Conradie puts it, "Our role is not to function as a second God, but to carry out a commission as a *primus inter pares* [first among equals] amongst those over whom we rule."[54] Or as H. Paul Santmire concisely and accurately puts it, we are in "partnership with nature."[55] The lands, airs, and seas, and the creatures over which we rule, are connected to us and we to them. As we have seen, our existence is constituted in our relationships with them. They depend on us, and we depend on them. We all are integral members of the ecosphere on whose proper functioning we all depend. If we thoughtlessly and unnecessarily alter or damage or destroy the ecosphere and its creatures, we are damaging and destroying ourselves. As the Apostle Paul wrote, "You reap whatever you sow" (Gal 6:7 NRSV). Our abuse of the earth and its creatures will eventually come around to hurt us—or our descendants.

(4) The Dominion Mandate Should Be Interpreted along with Genesis 2:15, the "Caretaking Mandate"

"The Lord God took the man and put him in the Garden of Eden to work it and take care of it" (Gen 2:15). In this passage, God places the human within his earthly creation (represented by the garden) for the purpose of working it and taking care of it. The word translated "work" is ʿābad, which literally means to serve. In this context it has the meaning of cultivating and nurturing the land like a farmer or a gardener. The word translated "take care of" is šāmar, which means to watch over, keep, preserve, or protect. Both words suggest action taken for the sake of the garden itself (the earthly creation), not for the gardener (us). This means that the dominion mandate has a caretaking, cultivating, and nurturing aspect to it. Thus, taking the dominion mandate together with the "caretaking mandate" of Genesis 2:15 gives balance to our role as stewards of creation. On the one hand, we are to use and modify the earthly creation for our needs, but, on the other hand, we are also to cultivate it, take care of it, and protect it for its own sake.

---

53. Bauckham, "First Steps," 236.
54. Conradie, *Ecological Christian Anthropology*, 204.
55. Santmire, "Partnership," 384.

(5) Dominion over the Earth Includes Dominion over Ourselves

The Apostle Peter counsels us to "make every effort" to develop *self-control* along with other virtues in our lives as Christians (2 Pet 1:6). Since we are part of creation, our responsibility to rule over it includes a responsibility to rule over ourselves. Not only are we to master and subdue the creation, we are to master and subdue ourselves—our own urges, appetites, impulses, self-interest, and drive for power, status, and wealth. History and our own experience tells us that self-rule is by far the most difficult aspect of the dominion mandate. It is one thing to master and subdue the sky with a jet plane, or a forest with a chainsaw, or a mountain with dynamite, or a river with a dam, or a bear with a rifle, but it is quite another to master and subdue ourselves. Our human desires and drives for food, wealth, and power must be moderated through the practice of conscious, prayerful, voluntary self-limitation and restraint. This is integral to healthy human life, to healthy dominion over God's creation, and to healthy ecosystems. As theologian Wolfhart Pannenburg notes, "Without self-rule there can be no unity or integrity of life."[56] In self-rule, we ensure not only the welfare of the lands, seas, and living things that God has placed under our control, but we also ensure the welfare of ourselves and our descendants. But, alas, a look at history tells us that, as theologian Hendrikus Berkhof put it, "we have been able to control nature . . . but we have been unable to control ourselves."[57] Consequently, as Susan Bratton wrote, "By not ruling our own appetites, we [will] ultimately cease to rule creation."[58] Human history is replete with examples of our failure to govern ourselves, and this is a major reason for the huge Ecological Problem we face today.

(6) Moderation—How Much Is Enough?

The debate about the dominion mandate sometimes seems to be between those who want virtually no human-caused changes in the earth and those who want unrestrained freedom for humans to do whatever they want. I think the better way lies between these two extremes. As

---

56. Pannenberg, *Systematic Theology*, 201.

57. Hendrikus Berkhof, *De mens onderweg. Een Christelijke mensbeschouwing*, cited in Conradie, *Ecological Christian Anthropology*, 203.

58. Bratton, *Six Billion*, 11.

I noted, we humans do need to modify parts of the earth and use its resources, but we should do it with moderation, gratitude, caution, and restraint. We need to rule over *and* cultivate, subdue *and* preserve.[59] But here lies the problem! Where is this middle way?! In the final analysis, our application of the dominion mandate is not a matter of doing this or that, up or down, black or white, all or none. It is a matter of how much, how big, how far, how fast, how powerful, how many, how long, how tall, how deep, how rich? It is a question of degree—when and where to stop. *When is enough enough?* We humans don't seem to do well with such questions. We seem predisposed to take things as far as circumstances, finances, and immediate conditions will allow. In the modern era, if not earlier, the paradigm has been "enough is never enough," or "bigger, richer, faster, more." Today, in the face of looming ecological limits, people may agree in principle on the idea of moderation and self-restraint, but they may disagree sharply on what this means for the human community and for each person individually. I will return to this difficult but vital issue in chapter 7.

(7) Dominion Is God's Blessing

Finally, we should recall that Genesis 1:28 is a blessing: "God blessed them [the humans] and said, 'Be fruitful and increase in number; fill the earth . . .'" God bestows upon human life the power of procreation, flourishing, and dispersal of human populations over the earth.[60] It is the same blessing and commission given to the sea creatures and the birds: "God blessed them and said, 'Be fruitful and increase in number and fill the water in the seas, and let birds increase on the earth'" (Gen 1:22). I suggest that neither the blessing of humans nor of the animals is to be interpreted literally nor as a mandate for unrestrained growth. The blessing of the animals is not confined to the birds and sea creatures but by implication is a general blessing for all the creatures of God's earth. Thus, the intent, it seems to me, when both the blessings are taken together, is for the flourishing and prosperity of God's creation as a whole—of all his creatures in interdependent community under the grateful, humble, self-controlled, cautious, and moderate dominion of

---

59. O'Donovan, *Resurrection*, 24–25.
60. Goldingay, *Old Testament Theology*, 1:54.

humans living in the image of God. The blessing is not a commission to engage in unrelenting and unlimited population growth for either humans or the animals. Neither is it a mandate for the unrestrained pursuit of wealth and affluence. This misconstrues the passage and confuses our creaturehood, which is characterized by limits, with divine existence, which is without limits. Ecologist Susan Bratton sees this in terms of evangelism. "The point has never been to produce biologically the most Christians, it has always been to bring the gospel and love of Christ to as large a proportion of the people as possible."[61] Similarly, Christian geneticist R. J. Berry argues that "the command to 'fill the earth' with our own descendants has become an evangelistic charge for spiritual increase through 'making disciples of all nations.'"[62] Whatever our interpretation of this blessing, we must understand that it is shared with all creatures and that it is limited.

Conclusion: We Are in Charge of God's Good Earth

We dusty earthlings are blessed with God's gift of dominion over his earthly creation. We are in control of the earth and its creatures. This is confirmed by science, theology, philosophy, and every other form of human inquiry. It is not up to dolphins, chimpanzees, wolves, or whales; it is up to us. We have to manage things. Just how we should do that, on what basis, to what end, and to what degree are the big questions. As I have said, our management should be characterized by gratitude, humility, caution, self-control, and moderation. I shall offer some further comments in this regard in chapter 7.

Dominion in the Image of Jesus Christ

The Bible's concept of rulership or dominion appears to embody the idea of ruling for the sake of the ruled. Several scholars have noted this. In contrast to other ancient Near Eastern concepts of rulership, the biblical rulers were supposed to rule for the sake of their subjects. Kings were to serve the people, not themselves (Deut 17:14–20). This reflects the character of God, who rules his creation and his creatures for their

---

61. Bratton, *Six Billion*, 131.
62. Berry, "One Lord, One World," 29.

Dusty Earthlings

own sake. Thus, our dominion function as God's representatives on earth would be to rule in the way of God.[63] If this mode of ruling is applied to the dominion mandate, then our rule should be carried out not for our own sake but for the sake of the nonhuman creation. Bauckham comments:

> Gn. 1:26, 28 gives human beings the status of rulers of the world, but it is not a biblical view of government that subjects exist for the sake of their rulers! If anything, the reverse is the case . . . Humanity's rule over nature is not intended to be tyranny, in which the ruler exploits his subjects for his own benefit, but good government, exercised responsibly for the good of the subjects. It is a share in God's rule over the world, and is therefore intended to reflect the fact that God does not exploit the world for his own benefit, but bestows his love and care on it because he values it.[64]

If this really is what the Bible means by human dominion over creation, then our actual exercise of this God-given dominion has been radically different from the biblical model. It's hard to imagine being more wrong.

Along the same lines, Bible scholar J. Richard Middleton finds a deeper meaning to the dominion mandate. In the gift of dominion, God graciously confers power on us his creatures so that we may be fruitful and multiply and fill the earth (Gen 1:28). According to Middleton, this is the divine pattern of rulership—to give his creatures space, freedom, and power to live, grow, and flourish. Middleton suggests that if we are to rule in a way that accurately images God's rule, then we should do the same thing for the creatures over whom we rule. In other words, just as the goal of God's rulership over us is our fruitfulness and flourishing by granting a measure of freedom and power to us, so also the goal of our rulership over other creatures ought to be their fruitfulness and flourishing by granting them a measure of freedom and power. After all, as we have noted, God's blessing fell upon the birds, fish, and animals as well as on us (Gen 1:22). The God we see in Genesis 1 is a "God who is generous with power, sharing it with creatures, that they might make their own contribution to the harmony and beauty of the world."[65]

63. Peacocke, *Creation*, 283.
64. Bauckham, "First Steps," 234.
65. Middleton, *Liberating Image*, 289.

Middleton concludes that the sort of power that humans are to exercise is a generous, loving power. It is power used to nurture, enhance, and empower others for their benefit, not for the self-aggrandizement of the one exercising power. "In its canonical place in the book of Genesis, the creation story in 1:1—2:3 thus serves as a normative limit and judgment on the violence that pervades the primeval history, indeed the rest of the Bible and human history generally."[66] Middleton points to the flood story in which Noah preserves and protects the animals as "an example of someone imitating the paradigmatic life-enhancing use of power that God is depicted as exercising in Genesis 1."[67] He adds, "Perhaps, then, our practice of reading (which we might call a hermeneutic of love) would be in harmony with the new ethic of interhuman relationships and ecological practice that we are aiming for and that is rooted in the *imago Dei*, an ethic characterized fundamentally by power *with* rather than power over."[68] Again, the stark contrast between Middleton's conception and our actual practice of dominion is evident.

Second, for us Christians, to rule in the image of God is to rule in the image of Christ, the true image of God.[69] In the gospel we encounter the idea of the good shepherd or the servant-ruler.[70] As theologian Wolfhart Pannenberg puts it, "When the royal rule of Christ is understood as the rule of the love of God that is revealed in the crucified Jesus, the customary notions of sovereignty among human beings are turned upside down."[71] In Christ, the ruler becomes the servant of the ruled. He rules for the sake of his subjects, even if there is cost to himself—a radical idea, indeed. Theologian F. LeRon Shults applies this to our rulership over creation: "If we are called to be like Jesus, then our 'rule' of the earth will take the form of servanthood, not oppressive domination over the creation of which we are an integral part."[72] Is Shults correct? Should this model of rulership be applied to our dominion over God's nonhuman creation? The notion that we are to be servant-rulers of wolverines and walruses, whales and wrens is indeed

66. Ibid., 295.
67. Ibid., 296.
68. Ibid., 297.
69. John 1:18; 1 Cor 4:4; Col 1:15; Heb 1:3.
70. John 10:11; Mark 9:35; 10:41–45.
71. Pannenberg, *Anthropology*, 76.
72. Shults, *Reforming Theological Anthropology*, 238.

strange to us. Given our normal attitudes and practices, applying this idea to the way we treat nature would involve a massive change to our worldview, attitudes, and practices.

If this is correct, or even partly correct, then for us who are redeemed and being transformed into the image of Christ (Rom 8:29), our way of living out the dominion mandate should, in tangible ways, imitate Jesus' conception of servant-leadership over creation. We should in some way be good shepherds in the way of Jesus. This would, it seems to me, motivate explicit support for and active participation in the preservation and restoration of ecosystems, restraint in the use of resources, protection of overexploited ocean fisheries, protection of endangered species and habitats, avoidance of damage to natural cycles (the carbon cycle, for example), the rescue of animals and seabirds caught in oil spills, and many other "creation care" activities. "For the New Testament authors, then, the concepts of the image and likeness of God refer directly to Jesus Christ, and our sharing in the glory of the divine image is possible only by our becoming like him and in him."[73]

Ecological Sin

In chapter 2 I reviewed historical, philosophical, and cultural factors that have led to our modern alienation from God's creation. Understanding these factors helps us see why we are where we are today, but the Bible holds that there is a deeper problem—our relationship with God's creation is broken at a spiritual level. The origin of this rupture is the fall of humanity described in Genesis 3. By this event, all our relationships were damaged—our relationships with God, with ourselves, with other humans, and with God's creation. The break between humans and the rest of creation is evinced by the exclusion (separation) of Adam and Eve from the garden (Gen 3:23–24). As we have noted, our relationality is fundamental to our being, so these broken relationships are hurtful to us. Taken together they constitute the basis of human sin and underlie the Ecological Problem as well as all other human problems. Most Christians I have talked to seem to agree that the Ecological Problem is due to human sin. Even conservative Christians who are often wary of ecological concern seem to agree that, if there is a problem,

---

73. Ibid., 239.

it is ultimately due to human sin. So what is sin, ecologically speaking? In what follows, I offer some thoughts that, I hope, will help contribute to an answer.

(1) Human Sinfulness Should Not Be Equated with Our Creatureliness

We humans are creatures—animals—and we do what animals do. We eat, drink, move, metabolize, build shelters, eliminate waste, reproduce, nurture our young, and so on. In so doing, we exploit other creatures and the ecosystems in which we live. These are not sinful activities. They are good activities that are part of who we are as eco-physical beings living in God's world. So to be an animal is not to sin. On the other hand, we are more than simply animals. We are human animals, ethical animals. Some sinful human activity may seem animal-like in that it resembles what we sometimes see other animals do (2 Pet 2:12). Selfishness, brutality, and excess may sometimes resemble what we see animals do in nature. But we are not obligated to live on this level. We have at least the potential to do better. We have a special relationship with God and are called to imitate God in the way we live in his world. This involves self-control, moderation, humility, justice, and love. If we behave in ways that are consistent with our animal life but not in terms of these higher human virtues, we may be sinning. Thus, to behave like a creature is not to sin, but to fail to behave like a human creature is to sin.

(2) Ecological Sin Can Be Manifested in Our Attempt to Replace God

The temptation of Adam and Eve in the garden consisted of the invitation to be "like God" in terms of knowledge and power (Gen 3:5). Based on this passage, theologian Millard Erickson says that the essence of sin is a "failure to let God be God. It is placing something else, anything else, in the supreme place which is his."[74] In our modern era, this "something else" is usually ourselves. Ecological sin can consist in our denial of our creaturehood and our dustiness, and the hubristic attempt to elevate ourselves to the position and role of God. But the biblical

---

74. Erickson, *Christian Theology*, 598.

view says that to be a human creature is neither to completely master nature nor to be mastered completely by it.[75] God gives us a certain amount of power (Gen 1:26–28) that is infinitely less than God's but greater than any other earthly creature's. Ecological sin can be manifested in our failure to understand the ecological limits that circumscribe our position and power within creation and our subsequent attempt (consciously or unconsciously) to transcend those limits. The Tower of Babel story in Genesis 11:1–9 is a biblical disquisition on this aspect of human sin. Our failure to take seriously our limitations, to use our power in moderation and restraint, to resist the urge to replace God, and to remember the lesson of the Tower of Babel are at the roots of ecological sin and the modern Ecological Problem.[76]

(3) Ecological Sin Can Be Expressed in Our Refusal to Accept Our Finitude

This concept is closely related to our attempt to replace God. Theologian Reinhold Niebuhr explained human sin in terms of a refusal to accept our finitude as human creatures. The problem is not our integration within the warp and woof of material, earthly nature but our prodigious efforts to escape it—to transcend our materiality and exceed the limitations of our created existence. As Niebuhr puts it: "Man is insecure and involved in natural contingency; he seeks to overcome his insecurity by a will-to-power which overreaches the limits of human creatureliness. Man is ignorant and involved in the limitations of a finite mind; but he pretends he is not limited. He assumes that he can gradually transcend finite limitations until his mind becomes identical with universal mind."[77] This effort to transcend our creatureliness and push the limits disrupts the order of creation. Spiritually it is expressed as rebellion against God. Socially it results in injustice.[78] Ecologically it results in the abuse and destruction of God's creatures and ecosystems. Although he was writing before the advent of the modern environmental movement, Niebuhr mentions the ecological aspect of this problem: "The

---

75. Hall, *Professing the Faith*, 340.
76. McGrath, *Re-Enchantment*, 79.
77. Niebuhr, *Nature and Destiny of Man*, 178–79.
78. Ibid., 179.

ego which falsely makes itself the centre of existence in its pride and will-to-power inevitably subordinates other life to its will and thus does injustice to other life."[79] Niebuhr helps us understand our sin in relation to the limits of our position within God's creation. He calls attention to our "ignorance of [our] ignorance,"[80] or our purposeful disregard of the fact that we do not know and will never know many things. Hence, humility—an awareness of our ignorance and our limitations—is a vital ecological virtue, an antidote to our inclination to overreach our finitude. As ethicist Larry Rasmussen puts it: "'Humility' means never trying to outgrow our 'humanity' and escape or transcend our earthiness. It means accepting ourselves for what we are—spirit-animated nature. To sin is to overstep and overshoot finitude, deny its potentialities and its limits, and reject creatureliness."[81]

(4) Ecological Sin Involves "an Unwillingness to Face Reality"[82]

Based on the Apostle Paul's discussion of sinful humanity in Romans 1:18–32, New Testament scholar Joel Green looks at our sin from another angle: "Paul recognizes that ungodliness and unrighteousness have as their object their own self-legitimation: humanity embraces a lie (1:25) and receives a corrupt mind (1:28), with the consequence that it defines its unjust ways as just. In other words, the conceptual patterns by which humanity perceives the world and orders its behavior is out of touch with the way things really are."[83] Human sin can involve our refusal to see things as they really are and live accordingly. If reality is too difficult, too inconvenient, or simply not the way we want it to be, we lie to ourselves, to others, and to God, and we construct alternative unrealities. Ecologically this is quite common today. We imagine ourselves to be exceptional beings, outside of creation, and exempt from ecological realities. We believe ourselves to be omnipotent, able to achieve any goal and conquer any problem. We believe we can ignore the ecosystem because whatever ecological problem may come along, our wondrous intelligence and technology will solve it. We may even

79. Ibid.
80. Ibid., 195.
81. Rasmussen, *Earth Community*, 275.
82. Erickson, *Christian Theology*, 633.
83. Green, *Body, Soul, and Human Life*, 102.

imagine that if this planet becomes too polluted or degraded, we will build spaceships and move to another planet somewhere. We construct an array of delusions to avoid the fact of our finitude. We live in denial. But as always happens with self-deception, reality sooner or later, in one way or another, imposes itself, and our delusions are smashed.

(5) "Human Sin Has Ecological Consequences"[84]

Since we are part of creation, since we are intimately connected to the ecosphere in which we live, and since we have dominion, what we do and how we live affect everything else. The Bible affirms this moral connection between humans and the ecosystem. The prototypical narrative for this is the flood (Gen 6:9—9:17).[85] "God said to Noah, 'I am going to put an end to all people, for the earth is filled with violence because of them. I am surely going to destroy both them and the earth'" (6:13). God destroyed all humans and "all life" with them (6:17). But along with Noah and his family, all species of animals were saved (6:19–21), and after the flood, God affirms human solidarity with the other creatures by making a covenant with Noah and "every living creature on earth" (9:10). "The reality is twofold—the animals perish with the people and the people and the animals are saved together. This is of the utmost significance for the history of humans and animals."[86] Humans and all living things live, die, and are saved together. "This new covenant [of Noah and the creatures] underscores the inherent relational connection of humans with the rest of creation. The Noahic covenant is a symbol of the unbreakable bond of all creatures with their Creator and with one another."[87] Humans and all other living things are bound together in both the consequences of sin and in God's deliverance from them. We are all in the same boat, as it were.

In the Hebrew prophets, we again see that human sin results in devastation of the land and its creatures. Jeremiah laments the damage to the lands of Palestine due to Israel's sin: "I will weep and wail for the mountains and take up a lament concerning the desert pastures. They are desolate and untraveled, and the lowing of cattle is not heard. The

---

84. Bauckham, *Bible and Ecology*, 94.
85. See also Hos 4:1–3; Jer 8:7; 18:14–16; Isa 24:1–4; Zeph 1:2–3.
86. Westermann, *Genesis 1–11*, 424.
87. Clifford, "Ecological Lament," 58.

birds have all fled and the animals are gone" (Jer 9:10). Jeremiah's lament here focuses on "the effects of sinful human choices on the land and the suffering of all creatures dependent on it for existence."[88] The prophet Hosea also directly connects ecological damage with human sin: "There is no faithfulness, no love, no acknowledgement of God in the land . . . Because of this the land mourns, and all who live in it waste away; the beasts of the field, the birds of the air and the fish of the sea are dying" (Hos 4:2–3).[89] Theologian William Dyrness sums up: "Morality, response to God, and fertility of the earth are interrelated . . . If there is any doubt about human responsibility for the earth and its preservation, these prophetic books should remove it. God is saying through these prophets that the very stability of the created order is dependent upon Israel's faithfulness to the covenant."[90] Our spirituality and our moral behavior are integrated with God's ecosystems. As we are seeing today, our sin impacts God's earth and his creatures in myriad ways.

(6) Human Sin Is Holistic

Sin, like salvation, involves our whole being and all of our relationships. Our relationships with God's creation and his creatures have not traditionally been included in our understanding of sin, but as we are learning, biblically, it is clearly there. It is important that we grasp this, not only because it is true but also because ecological sin is "eating away at the world we call home."[91] Theologian James Nash summarizes:

> Ecologically, sin is the refusal to act in the image of God, as responsible representatives who value and love the host of interdependent creatures in their ecosystems, which the Creator values and loves. It is injustice, the self-centered human inclination to defy God's covenant of justice by grasping more than our due . . . and thereby depriving other individuals, corporate bodies, nations, and species of their due. It is acting like the owner of creation with absolute property rights. Ecological sin is expressed in arrogant denial of the creaturely limitations imposed on human ingenuity and technology, a defiant disrespect

88. Ibid., 51.

89. For example, see Lev 18:24–28; Prov 2:20–22; Isa 24:3–6; Jer 4:23–28; 14:1–12; Ezek 36:1–12; Zeph 1:2–3.

90. Dyrness, "Stewardship," 57, 61.

91. Green, *Body, Soul, and Human Life*, 69.

Dusty Earthlings

> or a deficient respect for the interdependent relationships of all creatures and their environments established in the covenant of creation, and an anthropocentric [human-centered] abuse of what God has made for frugal use.[92]

We evangelicals have traditionally understood sin individualistically as one person's separation from God expressed in terms of personal disorder and moral failure. This is correct as far as it goes, but in reality sin is broader. It involves all relationships, including that between us and the rest of God's creation.[93] Our alienation from creation is manifested in many ways: our ignorance of and disregard for other creatures and the ecosystems in which we live; our economic system that ignores ecological realities; and our unjust social and economic arrangements by which some of us take far more than our share of God's earthly bounty, thus depriving other people and creatures of their due. But just as sin involves all of our being, individually and corporately, and all of God's earth, so also God's redemption involves all of this as well.

The Holistic Redemption of Christ

As I noted in chapter 1, many Christians understand salvation and eschatology in an individualistic and nonphysical way. God saves me, or my soul, out of this present, physical world, and, when I die, I will be transferred to a spiritual (nonmaterial) heaven where I will dwell with God forever. There are many variations of this, but for the most part the redeemed future of the faithful is spiritualized and individualized. We see salvation as for humans only, and we exclude the rest of creation—or perhaps we just don't think about it.[94] Whatever the specifics of this belief, it has had the effect of devaluing our bodies and the physical world, or of subordinating them to the overriding goal of going to heaven when we die. Jürgen Moltmann points out that this concept of salvation changes the meaning of the Lord's prayer: "The prayer for the coming of the kingdom 'on earth as it is in heaven' [is] replaced by the longing 'to go to heaven' oneself. The kingdom of God's glory and the salvation of the whole creation [is] reduced to heaven; and heaven [is] reduced

---

92. Nash, *Loving Nature*, 119.
93. Bookless, *Planetwise*, 37–38.
94. Berkouwer, *Return of Christ*, 212.

to the salvation of the soul."[95] This contracted notion of redemption, sometimes called "the immortality of the soul," has not always been the norm in the Christian tradition. Malcolm Jeeves observes that "the immortality of the soul is absent from the great creeds of the Christian church, for example, the Apostles' Creed and the Nicene Creed." These creeds do not speak of going to heaven when we die. "They speak of the resurrection of the dead and the life of the world to come. The emphasis is on resurrection, not on immortality."[96]

I suggest that a more biblical and theologically viable view holds that the redemption of Christ is, like the sin it remedies, *holistic*. It includes *all* of creation and us humans as part of it—including our bodies. Our great salvation hope is not to go to heaven when we die; it is the resurrection of the dead and the new creation.[97] As Paul writes in Colossians 1:19–20, Christ's sacrifice redeems "all things, whether things on earth or things in heaven," to God.[98] Many Bible scholars agree with this view. Richard Bauckham writes that Jesus' bodily resurrection "constitutes the redemption of human life in its psychosomatic wholeness . . . This holistic salvation of humans retains the solidarity of human physicality with the whole material creation, and cannot be conceived apart from the salvation of the latter too."[99] Theologian Terence Fretheim agrees: "Redemption is in the service of creation . . . . It is deliverance, not from the world, but to true life in the world. The redemptive victory of God frees the creation to become what God intended."[100] New Testament scholar Douglas Moo adds: "The resurrection of the body . . . necessarily requires an appropriate environment for [our] embodiment . . . Jesus' resurrection signals God's commitment to the material world."[101] Finally, New Testament scholar James D. G. Dunn helps explicate further Paul's conception of physical, holistic redemption and resurrection as physical transformation in 1 Corinthians 15:51–54. I quote him at length:

95. Moltmann, *God in Creation*, 181.

96. Jeeves, "Changing Portraits," 26.

97 Isa 65:17; 66:22; Matt 19:28; Acts 3:21; Rom 6:5; 8:19–22; 1 Cor 15:20–23; Rev 21:1–5.

98. The Greek word *ta panta* used in Colossians 1:20 is used in several places in the New Testament to denote all of creation, everything that exists apart from God. See, for example, John 1:3; Col 1:16–17; Rom 11:36; Heb 1:2; Rev 4:11.

99. Bauckham, *Bible and Ecology*, 171–72.

100. Fretheim, "Reclamation," 359.

101. Moo, "Nature in the New Creation," 462, 464.

> Redemption for Paul was not some kind of escape from bodily existence, but a transformation into a different kind of bodily existence . . . *Soma* [body] expresses for Paul the character of created humankind—that is, embodied existence. It is precisely as embodied, and by means of that embodiment, that the person participates in creation and functions as part of creation . . . [It] is what prevents the individual from opting out of this world or constructing a religion which denies social interdependence and responsibility . . . It is also this somatic character of Paul's anthropology which prevents Paul's theology from falling into any real dualism between creation and salvation. For it is precisely as part of creation and with creation that Paul the individual and his fellow believers share in the birth pangs of creation, groaning with the rest of creation as they await the redemption of their bodies (Rom. 8.22–23). In short, *soma* gives Paul's theology an unavoidably social and ecological dimension.[102]

Dunn ties together all the strands of holistic redemption and salvation in Paul's theology. Biblically speaking, we humans are physical beings, part of creation, and are redeemed and saved with and within God's whole creation.

At this point, I want to clarify two things about my argument for holistic redemption. First, from a biblical perspective, humans are the central objects of God's redemptive work, and the rest of creation is dependent on that.[103] As Paul writes, "For the creation was subjected to frustration . . . in hope that the creation itself will be liberated from its bondage to decay and brought into the glorious freedom of the children of God" (Rom 8:20–21). To claim that all of creation is redeemed through Christ is not to say that humans are removed from their preeminent place within creation. The redemption of the nonhuman cre-

---

102. Dunn, *Theology of Paul the Apostle*, 61. The biblically informed reader will recall that 1 Cor 15:50 says, "I declare to you, brothers, that flesh and blood cannot inherit the kingdom of God, nor does the perishable inherit the imperishable." At first glance, this verse seems to exclude the possibility that our resurrected state will be physical "flesh and blood." Bible scholar Joachim Jeremias published an article in 1956 in which he addressed this problem. He showed that Paul's term "spiritual body" does not mean that it is immaterial. Also, "flesh and blood," as Paul is using it here, means the non-transformed flesh and blood of this present age. What Paul is emphasizing is the radical transformation into a "different kind of existence," to use Dunn's words, that will occur at the resurrection when Jesus comes. Thus, like Jesus' resurrected body, which was physical yet transformed, ours will be the same (Jeremias, "Flesh and Blood," 151–59).

103. Gunton, *Christ and Creation*, 81.

ation is mediated through us humans. In his sacrifice Christ redeems us humans from our sins, but in so doing he also redeems the whole creation from its "bondage to decay."

Second, in making this claim, I am not proposing universal salvation for all humans. Within a doctrine of holistic redemption, there is room for the judgment of God.[104] This topic is beyond the scope of this book, but I want to make it clear that final moral and spiritual accountability before God for humans and for the nations is compatible with holistic redemption. Moreover, final accountability will include how we have treated God's earth and his creatures.

Finally, the process of God's holistic redemption has already begun with the coming of the Messiah Jesus Christ.[105] Jesus inaugurated the kingdom of God on his earth (Mark 1:15), and ever since his advent, the church in its life and ministry manifests (or should manifest) the kingdom of God and points toward (or should point toward) the final redemption of all things. This means that right now, today, the healing of the damage done to creation as a result of our sin and the healing of our relationships within it are taking place (or should be taking place). This healing should be visible in the way we, the redeemed people of the Lord, live in and treat God's creation. In other words, our ethics—how we actually live here and now—should manifest and point toward God's healing of ourselves and of his creation that was inaugurated in Christ's work two thousand years ago and will be consummated when Jesus comes again. As Francis Schaeffer expresses it, "The Christian who believes the Bible should be the man [or woman] who—with God's help and in the power of the Holy Spirit—is treating nature now in the direction of the way nature will be then. It will not now be perfect, but it must be substantial, or we have missed our calling."[106]

Likewise, in 2 Corinthians 5:17, the Apostle Paul writes, "So if anyone is in Christ, there is a new creation: everything old has passed away; see, everything has become new!" (NRSV). Traditionally, this verse has been understood individualistically as referring to the individual person and his or her new life in Christ. But Douglas Moo suggests that, as the NRSV translation indicates, this verse has a broader and deeper meaning. "This 'new creation' language refers to the entire new

---

104. 2 Cor 5:10; Heb 9:27; Rev 14:7.
105. Stassen and Gushee, *Kingdom Ethics*, 12.
106. Schaeffer, *Pollution*, 68–69.

state of affairs that Christ's coming has inaugurated . . . Roughly the same situation obtains in Gal 6:15, where 'new creation' is again used absolutely . . . While the phrase 'new creation' is not found in the [Old Testament], it is generally agreed that Paul's phrase ["new creation"] refers to the hope of a world-wide, even cosmic, renewal that is so widespread in the last part of Isaiah."[107] In other words, the worldview of the born-again Christian is transformed such that he or she *sees* the entire world in terms of the new creation in Christ Jesus.[108] As a result of this transformed worldview, his or her life and behavior *embody* this new perception of the world and foreshadow the healing and reconciliation of creation and all creatures in Christ. As Moo says, the idea that the world is destined for renewal at the return of our Lord should encourage "God's people to actively and vigorously align their values and behavior with what it is that God is planning to do."[109] This applies to all of life and all of our relationships, including our ecological relationships. Similarly, Bible scholar Peter Stuhlmacher argues that on the basis of our justification, regeneration, and renewal here and now in Christ, we Christian believers are transformed through the Spirit by the renewing of our minds (Rom 12:2) and are the "innovators of the good in the world" in advance of the coming of Christ in judgment. "Applied to the theology of creation, this means that *with their new orientation to life by faith in Christ, Christians are now well able to recognize God in his work in Creation again.*"[110] Of all people on earth, born-again Christians ought to be the most ecologically concerned and active of all!

Theocentrism

A major (perhaps the major) lesson we all have to learn as small children is that the world does not revolve around us: that everyone and everything are not there simply *for us*—we have to share, and we cannot always be the center of attention. I think any experienced parent knows that this is a challenging time as they gently move their youngster from a self-centered view to a more holistic view of their world. But for us

---

107. Moo, "Nature in the New Creation," 475.

108. This new creation is analogous to the "new heaven and new earth" that will come into being when Christ comes again (Isa 65:17; 66:22; Rev 21:1).

109. Moo, "Nature in the New Creation," 484.

110. Stuhlmacher, "Ecological Crisis," 10.

The Bible II: The Image of God, the Dominion Mandate, and Other Issues

Christian adults, both individually and corporately, we should take this further. The world does not revolve around us human adults either—neither socially nor ecologically. The created universe, including this earth, is not simply there *for us*. We are *not* the center of creation nor of this world nor of the ecosystem: *God is*. Simply put, *all things exist for the glory of God*.[111]

We Christians, mostly unconsciously, have retained a human-centered worldview in which we see the world in terms of ourselves. As a result, although we may be adults, our ecological attitude remains much like that of small children. We see creation and other creatures as there simply *for us*—for our needs and wants and little else. Creation exists as *our* "environment" that surrounds *us*. But biblically, creation is God's "environment"; it surrounds God, not us. God is at the center. This is what is meant by the word *theocentrism* (*theo-*, God, *-centrism*, centered): God-centered.[112] As the Apostle Paul writes, "Yet for us there is but one God, the Father, from whom all things came and for whom we live; and there is but one Lord, Jesus Christ, through whom all things came and through whom we live" (1 Cor 8:6).[113] The triune God is the center of the universe. Theocentrism emerges from a proper understanding of God as God and of ourselves as not God but as part of his creation. The seventeenth-century British naturalist John Ray (1627–1705) understood theocentrism and our difficulty with it. He writes, "It is generally received opinion that all this visible world was created for Man; that Man is the end [or goal] of the Creation, as if there were no other end of any creature but some way or other to be serviceable to man . . . yet, wise men nowadays think otherwise."[114]

Another biblical idea that leads directly to theocentrism is the notion that the purpose of all creation and all creatures is to praise God. "The creation exists for God's glory, not for human use."[115] Saint Francis of Assisi seems to have understood this well. His famous "Canticle of the Creatures" is decidedly theocentric. It begins:

111. Bauckham, *God and the Crisis of Freedom*, 153.

112. Rossi, "Theocentrism," 13.

113. See also Gen 1:1; Pss 24:1; 148; Matt 28:20; John 1:1; Rom 11:36; Eph 4:4–6; Col 1:15–20; Rev 1:8; 5:11–14.

114. John Ray, *The Wisdom of God Manifested in the Works of Creation*, quoted in Nash, *Rights of Nature*, 21.

115. Bauckham, *Living with Other Creatures*, 163.

Dusty Earthlings

> Most high, omnipotent, good Lord,
> All praise, glory, honor, and blessing are yours.
> To you alone, Most High, do they belong,
> and no man is worthy to pronounce your name.[116]

This aspect of Francis's worldview bore fruit in the way he lived simply, humbly, kindly, and respectfully among God's creatures. The notion that "the whole earth is full of [God's] glory" (Isa 6:3) leads directly to theocentrism—all exists for the sake of God.

The view that all things exist for our sake is called *anthropocentrism* (*anthropo-*, human, *-centrism*, centered). It denotes the view that the earth, the biosphere, all creatures, and the entire creation are all there for the sake of us humans. This is by far the dominant view in the world today, and it is the view that most of us Christians hold and live out in our lives. It is so fundamental to our perception, thinking, feeling, and life that we are unaware of it. It's just the way things are. But the Bible sees the world differently. It presents God not only as the author, sustainer, and redeemer of all creation, but also as its centering focus. This is depicted in Figure 1.

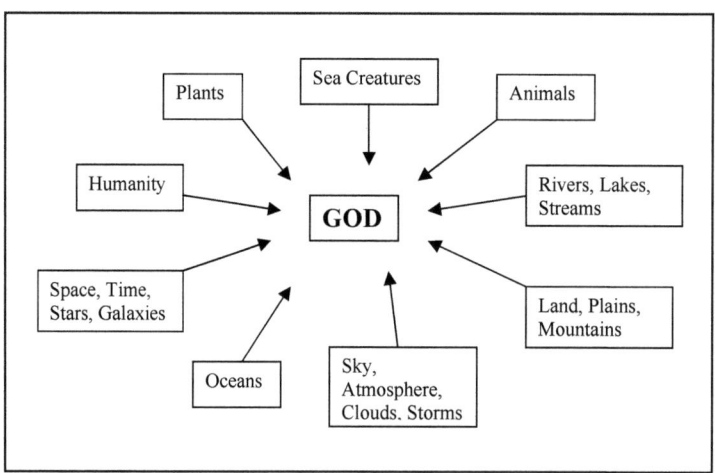

Figure 1. Theocentrism: everything, including humanity, is or should be centered on God.

While we are the creatures to whom God has given dominion over his creation, this does not abrogate creation's theocentrism; it enhances it. If God has made us managers of his creation, he has done so to bring

116. Sorrell, *St. Francis*, 101.

glory to himself, not to us. We are to carry out our role as the managers of nature to the glory of God. In our dominion we should be leading the rest of creation in focusing on God and praising God. David Gushee writes: "*What makes creaturely life sacred is God's relation to it, not any particular characteristic we might claim for ourselves or any other creature . . .* All creatures bow before the majestic Creator who alone gives them value."[117] The Apostle Paul provides us with a beautiful and succinct statement of the doctrine of theocentrism: "For from him and through him and to him are all things. To him be the glory forever" (Rom 11:36).

A theocentric worldview implies that the world may not always align itself with our human goals and preferences. Indeed, the world as we find it seems to manifest this reality. Ethicist James Gustafson challenges us: "What we judge to be good for man, or for a human person, or some human group, may not be in accord with the ordering purposes of God, insofar as they can be discerned . . . It may be that the task of ethics is to discern the will of God—a will larger and more comprehensive than an intention for the salvation and well-being of our species, and certainly of individual members of our species."[118] If all things are centered on God, then our ethical task is to align ourselves, as best we can, with God's larger purposes. Philosopher Stephen Toulmin comments further: "For anyone who adopts a truly *theocentric* ethics, the basic moral imperative is, rather, 'to conduct life so as *to relate to all things in a manner appropriate to their relations to God*.'"[119] This book is an attempt to move in that direction (Matt 6:10).

While the world exists for God in a primary sense, it also exists in a secondary sense for us and for all living things. All organisms require the support of their habitats and ecosystems to survive and thrive, and we humans do too. Thus to say, for example, that the land yields its crops or the tree bears fruit that we humans might be fed and thrive is correct, as long as we remember that the primary goal of the land's crops and the tree's fruit—and of everything else—is to glorify God.

Theocentrism is very distant from our current worldview as modern Christians, especially in things ecological. It is extremely hard for us to grasp, let alone implement. How to approach and live out this doctrine has not been well articulated by Christian theologians and

---

117. Gushee, "Environmental Ethics," 264.
118. Gustafson, *Ethics*, 113.
119. Toulmin, "Nature and Nature's God," 39.

ethicists—perhaps because, as Gustafson recognizes, it is so challenging to our status quo. But if we are to become theocentric people, we must begin the process. We have much work to do.

Conclusion: It's About God

We have covered a lot of ground in these two chapters. In chapter 4, we found that God is the one and only self-existent being who is separate from and above creation (transcendence) but present in and directly related to it (immanence). He is the creator, owner, provider and redeemer of all that is, and he delights in what he has made. He is a kenotic God of self-giving love toward his creation. By contrast, we humans are creatures; we are not God. We are dusty earthlings—physical, contingent, dependent, limited beings, organically embedded within God's earthly ecosphere. We noted the fact that Jesus Christ, the incarnate God, is the archetype of embodied humanness. The basic thrust of chapter 4 was to let God be God and to place us humans firmly where we belong—within God's creation.

In this chapter, we looked first at humans as the image of God. We concluded that the *imago Dei* sets us apart from other creatures but that it must be understood in terms of our eco-physical nature as creatures—part of God's creation. Ultimately, the *imago* is embodied paradigmatically in Jesus Christ, the one toward whom we humans aspire in our worldview, attitudes, and life. We then looked at the dominion mandate. In granting us dominion, God endows us humans with power and responsibility over his earthly creation, but this power carries with it certain conditions: limitedness, responsibility, and giftedness. We are to use our dominion to care for creation, not merely to exploit it (Gen 2:15). Since we are insider stewards, part of creation, dominion is not only about ruling and subduing the nonhuman creation, but, more importantly, it is about ruling and subduing ourselves (voluntary self-control). This is related to the kenosis of God in Jesus Christ, who in his incarnation and death voluntarily limited himself out of love for us and for his creation. Finally, I suggested that the passage "Be fruitful and increase in number . . ." (Gen 1:22, 28) should be understood as a blessing for the prosperity and flourishing of all of creation as a whole, not just humans.

The Bible II: The Image of God, the Dominion Mandate, and Other Issues

We then discussed ecological sin. In a primordial sense, sin involves rupture of all our relationships including our relationship with the nonhuman creation. We noted that from an eco-physical perspective, sin is a refusal to accept our creaturehood, a denial of our finitude, and an effort to transcend the limits of our created existence—in effect, an attempt to become God. Sin is pervasive in that it disrupts the entire fabric of God's ecosphere.

God's redemption is holistic in that it remedies the sin problem in all its aspects, including the ecological. Thus, our faith is a resurrection faith, a *this-worldly* faith of renewal and transformation here and now on this earth. We look forward, not to going to heaven when we die, but to the resurrection of the dead and the renewal of all things on that great day of the Lord when Jesus comes again. The advent and work of Christ—his birth, life, teaching, miracles, death, and resurrection—inaugurated the kingdom of God on earth, in which all of his followers seek to live, thus foreshadowing the ultimate healing and renewal of God's earth in the coming of our Lord.

Finally, we discussed theocentrism, the biblical idea that God is the center and focus of all of creation. All that God has made exists to his glory and for his purposes. The thesis of this book—that we are a part of creation, not separate from it—means that we have no separate existence independent of the broader creation. In this sense we are not special. We too take our place among God's creatures as part of God's "environment" of praise and adoration.

At this point, I hope I have established the first two arguments of this book. First, we are physical beings, dusty earthlings, and second, as physical beings, we are necessarily ecological beings, part of the earth's ecosystem. My third argument is that as eco-physical beings, we are subject to all the patterns, principles, parameters, and limits (ecological realities) that govern the existence of all creatures on this planet. In chapter 6, I shall discuss more fully what some of these ecological realities are and what they might mean for how we live on God's good earth.

6

Ecology for Dusty Earthlings

As ECO-PHYSICAL BEINGS LIVING in God's eco-physical world, we are subject to all the ecological realities that govern the existence of all things living on earth (ch. 1, page 4, argument 3). The dominion mandate says that we are to manage creation in such a way that all humans and all creatures can live together and flourish in God's world. This can only happen if we manage creation, including ourselves, in a manner aligned with these ecological realities. The ecosphere consists of a complex system of "checks and balances, controls and feedback loops," energy and resource exchanges, and other relationships and interactions that support us and that circumscribe our existence.[1] In what follows I shall discuss some of these and suggest some of their implications for our lives.

The Science of Ecology

Ecology is the study of living things and their relationships and interactions with one another and with their nonliving environments. As a science, ecology exhibits some peculiarities that perhaps make it unique among the natural sciences. I want to mention a few relevant characteristics of ecology before I get into the substance of the chapter.

(1) Ecology Is a Young Science

In contrast to, say, physics or astronomy, ecology has only been around for about a hundred years or so. It has made great progress in that time

---

1. Rolston, "Kenosis and Nature," 51, 53–54.

but is still in some respects a young science. Despite ecology's impressive progress, some scientists worry that it may not be advancing fast enough to keep up with the massive impact that humans are having on the global ecosphere and on local ecosystems.[2] Like the teenage student I mentioned in chapter 1, we are rushing ahead, not really knowing where we are going. Ecologists are working hard to provide us with information to help us see where we are going, but it is an open question whether they will be fast enough to provide adequate information (especially in the eyes of skeptics) or whether we will blunder into disaster or something else before knowledge, wisdom, and humility are acquired.

(2) Ecology Is an Interdisciplinary Science

Ecology in its broader applications brings together many sciences and fields of study including hydrology, soil science, geology, climatology, physics, chemistry, botany, zoology, taxonomy, genetics, and more. In fact, since humans are part of the ecosphere (as I am arguing in this book), the human sciences—history, neuroscience, psychology, sociology, political science, economics, philosophy, theology, ethics, and so on—can be part of the paradigm in ecological study. In our modern age of specialization, the interdisciplinary nature of ecology has sometimes generated controversy. But in recent decades, there has been a trend toward interdisciplinary exchange in many fields, so ecology has, perhaps, felt vindicated. For us laypeople, it is well to remember this interdisciplinary nature of ecology and the complexities it involves.

(3) Ecological Systems Are Open, Dynamic, Complex, Nonlinear Systems

Ecosystems consist of an array of organisms living together in complex relationships in a given area or region. All ecosystems are *open* in that they receive energy input each day from the sun,[3] and they exchange energy, matter, information, and organisms with surrounding ecosystems.

---

2. Belovsky et al., "Ten Suggestions," 346.

3. This is not true for deep ocean ecosystems that exist in complete darkness. These systems receive energy input from a steady rain of organic matter descending slowly from shallower depths or from the mineral rich waters of geothermal vents.

They are not closed or isolated from the broader ecosphere. They are dynamic, meaning that they are constantly changing. Ecosystems are *complex*: they involve multiple organisms, relationships, processes, and factors, all interacting in dynamic ways over space and time. A desert landscape, for instance, may appear to be relatively simple—just sand and rocks—when in reality it contains thousands of organisms, interactions, and relationships, all continuously changing. Ecosystems are *nonlinear*, meaning that there is almost never a simple, linear, causal chain of events where A leads to B leads to C, and so on. There is almost always a network of events with multiple factors interacting in a variable, three-dimensional matrix of feedbacks, feed-forwards, loops, exchanges, checks, balances, and so on. Ecologists can isolate some of these interactions and gain information, but they have to take into account other factors that are impinging on the particular part of the system they are studying. For example, an ecologist may study changes in a squirrel population in a given ecosystem and find that they are related to predation by a species of hawk. But there are other factors at play such as food availability, climate, competition, disease, female fertility, immigration and emigration, human activity, and so on. As you read this chapter, keep in mind the open, dynamic, complex, and nonlinear nature of earthly ecological life.

(4) Ecological Knowledge Is Uncertain and Probabilistic, but So Is All Human Knowledge

The global ecosphere in which we live is incredibly complex, surpassed, perhaps, only by the human brain. But again, as I am arguing, the seven billion human brains (or rather brain-bodies) inhabiting this world are also part of the ecosphere. Thus, if we include humans, the earth's ecosphere is by far the most complex entity we know of in all of God's creation. The famous biologist E. O. Wilson describes the ecosphere as "a stupendously complex layer [covering the earth's surface] of living creatures whose activities are locked together in precise but tenuous global cycles of energy and transformed organic matter. The biosphere creates our special world anew every day, every minute, and holds it in a unique, shimmering, physical disequilibrium."[4] Wilson's description

---

4. Wilson, *Future*, 39.

is flowery but helpful. The biosphere is indeed a vast, dynamic, highly complex system. It is not a deterministic clockwork machine that follows "laws" in an absolutely predictable fashion. As we will see, this "shimmering physical disequilibrium"[5] is that on which we humans and all of God's creatures depend for our lives. Because of its complexity, we find that our knowledge of it is uncertain and our predictions of its behavior are always probabilistic.

But uncertain and probabilistic knowledge is not unique to ecology. Much of human life and knowledge is like that. Every day we make choices and act on the basis of partial, uncertain, and probabilistic knowledge. A young man in love asks a woman to marry him thinking she would *probably* accept. We invest in the stock market or operate a business based on *uncertain* knowledge about the market and the economy. This is simply the nature of our existence as dusty earthlings on this earth as God created it. As we saw in chapter 4, God knows all things perfectly; we don't. And ecology is no exception.[6]

Having said this, I need to add that we do have a great deal of ecological knowledge that, while not certain, is good enough to act on.[7] Our ecological knowledge, like our biblical knowledge, may be partial and uncertain, but that is not the problem. We know enough to know what we ought to do—especially when ecological knowledge is coupled with biblical knowledge. Our problem is not knowledge; it is life and action. We have knowledge, but we do not have the moral and spiritual

---

5. The "disequilibrium" of which Wilson speaks here is an important concept. See the section on the *new ecology* at the end of this chapter and in the glossary.

6. In the global climate change debate, skeptics sometimes argue that our scientific knowledge of climate change is "unproven" and we should not act on it until it is "scientifically certain." As I have noted, most human knowledge, including ecological knowledge, is inherently uncertain and always will be. Some of these very same folks who demand "proof" and "certainty" before acting on climate change invest in the stock market, putting their hard-earned money at risk in a highly probabilistic and uncertain phenomenon. If there is anything in the world that is uncertain, surely it is the stock market. In point of fact, if we are going to demand proof and certainty before we do anything, then we will never do anything because proof and certainty will never come. There are legitimate issues to debate regarding climate change, but this is not one of them. It should be set aside. For a forthright and accessible discussion of the uncertainty of human epistemological life, see Taylor, *The Myth of Certainty*. He comments, "My own experience is that for human beings certainty does not exist, has never existed, will not—in our finite states—ever exist, and, moreover, should not. It is not a gift God has chosen to give His creatures, doubtlessly wisely" (94).

7. Orr, "Retrospect and Prospect," 1350.

Dusty Earthlings

resolve to heed that knowledge and live as we ought to live and do what we ought to do. I shall return to this in chapter 7. For now, let's move on to the substance of this chapter: what are some of the ecological realities that define and circumscribe our existence as dusty earthlings living on planet earth?

Some Principles, Patterns, Parameters, and Limits of Earthly Eco-Physical Life

Feedback Loops

Feedback loops are important ways in which different parts of ecosystems interact with one another. Ecosystems and the entire ecosphere are replete with networks of positive and negative feedback loops. Feedback means that a change in one part of the system feeds back to another part that in turn affects the first part. The form of the feedback can be energy, material, or information,[8] and can occur in either a positive or negative way. *Positive feedback* means that an increase in one part of the system feeds back to *increase* another part that in turn *increases* the first. *Negative feedback* means that an increase in one factor feeds back to *decrease* another part that in turn *decreases* the first. Some examples will make this clear.

The relationship between populations of a predator and its prey can behave in the manner of negative feedback loops. Predators are animals like hawks, mountain lions, and large fish that feed or prey on other animals like field mice, squirrels, and smaller fish. An example is a negative feedback loop that appears to play a role in population fluctuations that have been observed in snowshoe hares and Canadian lynx in the northern forests of Canada and Alaska. Close relatives of jackrabbits, snowshoe hares, as their name implies, have large feet that are covered with fur so that they stay warm and don't sink into the snow in the winter. Canadian lynx are medium-sized wildcats that prey on snowshoe hares. Serendipitously, since about 1800, the Hudson Bay Company of Canada kept meticulous records of the trapping and killing of lynx and hare so that today ecologists can estimate population

---

8. Odum, "New Ecology," 15.

Ecology for Dusty Earthlings

sizes and trace their ups and downs over the last two hundred years. It turns out that these two populations cycle through dramatic boom and bust phases every ten years or so. Extensive studies have been done to determine the cause(s) of these cycles. Although there are multiple factors involved, negative feedback interactions between lynx and hares appear to play an important role.[9] When hare populations are low, there is less food available to the lynx, and this feeds back negatively to reduce their population. Subsequently, with lower predation by the lynx, the hare population begins to increase again, which in turn feeds back positively to the lynx by providing increased food and an increase in their population. In fact, when hare populations are at their peak, the lynx actually engage in "surplus killing," taking more hares than they need and leaving them uneaten. In other words, they waste resources.[10] (Like the lynx, we humans also exhibit this same behavior, overexploiting and wasting resources in the face of abundance.) Again, the dynamics of these cyclic variations in these wild populations are complex, but we can see that they are linked by a negative feedback loop involving the availability of food and rates of exploitation that regulate their respective numbers.[11]

In humans, small-scale human societies often contain negative social feedbacks such as taboos or moral proscriptions under certain conditions that suppress behavior that may be detrimental to long-term resource sustainability. For example, some aboriginal societies have had taboos against excessive hunting. The taking of a certain number of animals or other conditions triggered socially enforced prohibition of further hunting, thus preserving the prey population and helping ensure a long-term resource. When these societies are incorporated into larger industrial cultures, these social feedback loops tend to disappear, resulting in overexploitation and unsustainable behavior.[12] But similar negative feedbacks have also emerged in industrial societies. In the face of overhunting and overfishing, we have established regulations and agencies, licensing, limits, and other controls that form a negative feedback loop and help minimize overexploitation of prey populations.

Modern farming presents an example of a positive feedback loop in which technology leads to more technology. Today farming involves

9. Krebs et al., "What Drives," 32.
10. Ibid., 29.
11. Molles, *Ecology*, 331–34.
12. Bernstein, "Ecology and Economics," 325.

159

the exclusion of as many species of plants and animals as possible from the ecosystem and the installation of a monoculture (single species) of, say, wheat, corn, or soybeans. With the suppression of biodiversity, the normal feedbacks that replenish the soil, maintain productivity, and control weeds and pests are eliminated. Herbicides and insecticides are required to control these factors. These chemicals induce resistance in weeds and pests that then requires the development of new herbicides and pesticides that leads to the emergence of more resistance and so on. Loss of ground cover leads to increased nutrient loss by leaching and erosion, which requires increased applications of fertilizers. Increased runoff to rivers leads to increased flooding downstream, which requires the engineering of rivers through the construction of dams, channels, levees, pumps, diversions, and so on. Technology feeds back to produce more technology. This is a complex positive feedback loop that is extremely common in the modern world.

Another example of a technological feedback loop is our modern food production and the epidemic of overweightness and obesity that it has produced in human populations and our response to that epidemic. As a result of advances in farming, food processing, distribution, government policies, marketing, and advertising, we have, in many human societies, an overabundance of high-calorie, low-quality food. At the same time, improvements in transportation and other labor-saving devices have reduced the demand for exercise and calorie expenditure. As a result of these technologies, people are taking in more calories and expending less, resulting in widespread overweightness and obesity in the population. This has fed back into the system to generate more technology in the form of diet programs, drugs, bariatric surgery, expanding medical technology to treat the complications of obesity, and an exercise industry with fitness centers, machines, personal trainers, and programs designed to induce humans to exercise. In a positive feedback loop we produce machines and technology to save calories, then produce more machines and technology to expend the calories we saved. Common sense would suggest that rather than developing all this costly technology to deal with the consequences of technology, we merely reduce the food supply and reduce our use of labor-saving devices. This would improve the health of the population and save money and energy without adding more layers of technology. This, in effect, would convert a positive feedback to a negative one. But due to certain cultural and economic concerns and values we do not do that.

As I noted, this more-technology-yields-more-technology positive feedback loop is very common in modern human society. Although it may be economically beneficial in terms of sustaining existing markets, producing new markets, and generating jobs and monetary wealth, it may not be ecologically beneficial or sustainable in the long term, not to mention the damage it does to human health and well-being.

An important aspect of ecological feedback is *time lag*. That is, there may be a delay in feedback such that one factor continues to increase for a time before it begins to feed back on the factor that is increasing it. When this occurs, there is a tendency for more erratic and exaggerated ups and down in both factors. The dramatic booms and busts in the snowshoe hare and lynx populations in Canada and Alaska are thought to be due, in part, to a time lag in feedback between the two populations. Another example might be human carbon production and atmospheric temperatures. Because of the buffering effect of the world's oceans in their capacity to absorb large amounts of both CO_2 and heat, as well as other factors, humans can produce a lot of CO_2 before they notice a clear-cut rise in temperatures and begin decreasing their CO_2 output. There is a time lag between CO_2 production and actual temperature rise and climatic and ecosystem changes, during which we continue to produce CO_2. As a result, there could be an "overshoot" such that atmospheric CO_2 concentration and temperatures rise higher than they would if we humans received earlier feedback with less time lag.

In today's globalized world, the complex worldwide exchange of materials, energy, resources, and information has a buffering effect that generally increases the time lag of feedback loops regarding the use of energy and material resources in any given region or ecosystem. For example, here in Southern California, where I live, the local human population far exceeds the region's capacity to produce food. But since we have the technology, money, and energy resources (fossil fuel) to import food, the feedback of the ecosystem to limit our food supply is muted (for the time being). As long as we are able to continue importing large quantities of food, we will be shielded from the effects of our local overpopulation. Time lags in feedback loops can make ecological planning challenging and controversial.

Since the beginning of the modern era, we have been more or less committed to a command-and-control approach, seeking to overpower feedbacks and control natural processes, and thanks in no small measure to massive inputs of energy from fossil fuels, we have had some success doing this. This in turn has given us the illusion that we are exempt from ecological feedbacks. But this is not so. Among other things, our industrialized consumer culture does not take into account the simple negative feedback loops of ecological supply and demand. We are currently overexploiting many ecological resources and many ecosystems are being degraded.[13] We are living beyond our means. Like the lynx that overexploit their food supply in times of abundance, not noticing feedback time lags, thus suffering population crashes, we humans are overexploiting resources (fossil fuels, farmland, ocean fisheries, ecospheric CO_2 absorption capacity, etc.), failing to notice feedbacks and time lags. As a result, we may find ourselves facing serious problems in the future—like the crashes of the lynx and snowshoe hares.

Resource Recycling

God endowed his earth from its beginning with finite amounts of the resources needed to support life and biodiversity. The fact that billions of plants, animals, and humans have lived and died on the planet since creation without running out of these resources, is due to an array of recycling systems within the ecosphere. As Christian biologist Calvin DeWitt writes, "Cycles upon cycles . . . cycles within cycles . . . cycles of cycles . . . the creation is permeated with cycles."[14] Well understood ecological cycles include those involving water, carbon, nitrogen, sulfur, oxygen, and phosphorus.[15] These cycles are complex pathways by which these resources are converted from one form to another, exchanged between sinks (storage reservoirs within the ecosystem), and then passed on to other components of the system in a cyclic manner. In this way, these resources, and many others, are recycled over and over again, allowing sustainable habitation of the planet over time. Energy for resource recycling comes from the sun, and recycling constitutes a major part of God's providence for us and for all his creatures on earth (Gen 1:14–18; 8:22; Ps 104:10–23).[16]

Take the *nitrogen cycle*, for instance. We, like all animals, depend on the nitrogen cycle for our lives. Nitrogen is an essential ingredient in

13. Costanza et al., "Value of the World's Ecosystem," 259.
14. DeWitt, *Earthwise*, 18.
15. Campbell and Reece, *Biology*, 1196–97.
16. DeWitt, *Earthwise*, 17.

Ecology for Dusty Earthlings

proteins, nucleic acids, and other vital biological molecules in all living things, including us humans. Nucleic acids form the DNA that makes up our genes. Proteins support the structure of our bodies, serve as hormones transmitting information around our bodies, form enzymes that catalyze the millions of chemical reactions on which our lives depend, and form the tough, cross-linked polymer called keratin that makes up our skin and hair.

The atmosphere is about 78 percent nitrogen gas in its elemental form (N2) that is not biologically available to most organisms including us. But there are bacteria living in the soil and in the roots of plants called legumes that can absorb and convert atmospheric nitrogen gas into ammonia and nitrates that plants can use.[17] This process is called nitrogen fixation (N fixation). Plants incorporate the nitrates into their tissues, which we and other animals then eat—or which other animals eat, and then we eat those animals. In this way, the ecosystem provides us and all the creatures with the nitrogen we need. We and other animals then excrete excess nitrogen in our urine, or when we die our bodies decompose, returning the nitrogen to the soil. Putrefying bacteria and fungi recover the nitrogen from our urine and decaying bodies and convert it to a form in the soil that plants or other bacteria can use. At the same time denitrifying bacteria are continually taking up soil nitrogen, converting it back to N2 and returning it to the atmosphere to complete the cycle.[18] The nitrogen cycle is complex, but this is the basic idea. It is an amazing system—a manifestation of God's loving providence for us and for all his creatures.

Ecologist Peter Vitousek and others have studied the nitrogen cycle and found that in the modern era, we humans have radically altered it. Like the nitrogen-fixing bacteria in the soil, we take nitrogen out of the atmosphere and produce ammonia and many nitrogen-containing products such as fertilizers.[19] Through our use of these products, these chemi-

---

17. Soybeans, peas, beans, peanuts, alfalfa, and certain other plants are called legumes. They have nodules on their roots containing N-fixing bacteria that allow them to flourish in soils low in nitrogen. Farmers may take advantage of this by using crop rotation, alternating legumes with, say, corn or wheat. The legumes replenish the nitrogen in the soil and reduce fertilizer requirements.

18. Odum and Barrett, *Fundamentals*, 143–48.

19. One major way that we do this is through what is called the *Haber process* (named in honor of the German chemist Fritz Haber, who developed it in 1905). In this process N2 is removed from the atmosphere and converted to ammonia, which is

cals reenter the system with the effect of accelerating the nitrogen cycle by increasing the transfer of nitrogen from the atmosphere to the soil and to other parts of the ecosphere such as lakes, rivers, and oceans. Also, during the combustion of fossil fuels (in our cars, for example), atmospheric nitrogen is converted to biologically active nitrogen oxides and ammonia and released into the atmosphere in auto exhaust (and from other sources).[20] This eventually ends up back in the soil and in water sources. Cultivation of nitrogen-fixing crops such as soybeans also fixes even more nitrogen. These factors have at least doubled the global rate of N fixation and increased the bio-available forms of nitrogen in soil and water sources. We are also increasing the amount of nitrogen compounds in the atmosphere (N_2O, NO_2, NO, NH_3).[21] Although we don't know all the long-term effects of this, we do know that it is contributing to a phenomenon called *eutrophication* in lakes, streams, and certain parts of the oceans.[22] This can reduce the ecological productivity and biodiversity in these places. When nitrogen oxides dissolve in water, they form acids that can be detrimental to life.[23] Vitousek says that we are "certain" that these changes exist and that they are human-caused.[24]

The changes we are making in the global nitrogen cycle are unintended. We did not consciously plan for them, nor are we consciously aware (except for people like Vitousek) of what the ultimate effects might be. Vitousek says these changes in the nitrogen cycle plus increasing concentrations of carbon in the atmosphere and changes in land cover are all firm evidence that we humans are in the process of

---

then converted into nitrates for fertilizers, explosives, household cleaners, and much more.

20. Nitrogen compounds such as nitrogen dioxide (NO_2) are major components of air pollution and smog. Many devices have been developed to reduce the output of nitrogen-containing substances into the atmosphere. Catalytic converters in our cars and scrubbers in coal-fired power plants, for example, are designed to remove these substances, along with other pollutants, from their emissions.

21. These compounds are also greenhouse gases.

22. Eutrophication is a process whereby nutrients (mainly phosphorus but also nitrogen compounds) are added to the water, leading to an overgrowth of algae. When the algae die, they sink and are decomposed by bacteria and other organisms. This removes oxygen from the deeper waters. Without oxygen, fish and other larger animals cannot survive. The end result is a body of water filled with algae, without fish or other animals that require oxygen.

23. Vitousek et al., "Human Alteration," 738–39.

24. Vitousek, "Beyond Global Warming," 1865.

unintentionally altering the entire global ecosystem.[25] "It is clear that the current explosion of human activity is changing the way the world works, not just through an increasing exercise of dominion and control over nature, but also because the scale of our activity is so great that even our waste products loom large globally."[26]

The *carbon cycle* is essential for life. Green plants, algae, phytoplankton, and photosynthetic bacteria remove from the air the carbon dioxide (CO_2) that we and other animals breathe out. Using energy from the sun, they convert that CO_2 to organic carbon in the form of sugars, starches, fats, and proteins, and incorporate it into their stems, leaves, seeds, and fruits. We eat it or feed it to other animals that we then eat. In this way we receive the carbon we need. We are absolutely dependent upon green plants, algae, photosynthetic bacteria, and phytoplankton for the cycling of carbon in the ecosphere. They also supply the oxygen we require to live. Without them, the earth would be uninhabitable by oxygen-loving organisms like ourselves and most all other animals. Today we are altering the global carbon cycle. By burning fossil fuel at high rates, we are increasing the concentration of CO_2 in the atmosphere. Since CO_2 helps the atmosphere retain heat, its rising concentration is slowly raising average global temperatures and changing the climate. The consequences of this are difficult to predict, but there is evidence that they may be detrimental to humans, especially poor and vulnerable humans, and to many other living things.[27]

In the carbon cycle, carbon is cycled among several sinks (storage reservoirs) including the atmosphere, oceans, soil, rocks, vegetation, and animals. Tropical rainforests, such as those in the Amazon basin of South America, in central Africa, and in Southeast Asia, store almost all their carbon and nutrients in the living substance of trees and vegetation; little gets stored in the soil since the high rainfall tends to leech away nutrients from the soil into streams and lakes, where it is lost from the system. Thus, when the trees and vegetation are removed, the main carbon sink for the ecosystem is lost, and the soil is found to be of poor quality for farming. Soil exhaustion has been one of the causes behind the destruction of rainforests. As the soil is exhausted in one location, farmers who may not be able to purchase fertilizers to replenish the soil

25. Ibid., 1862.
26. Ibid., 1872.
27. See Houghton, *Global Warming*.

Dusty Earthlings

may find it easier and cheaper to move to new forestland, remove the trees and start again. This situation contrasts with temperate forests, such as those that cover much of eastern North America, where rainfall amounts are lower and leeching is much less of a problem. Here most of the carbon is stored in the soil. Removal of the trees and vegetation does not remove the primary carbon sink, and the soil is found to be quite fertile for farming.[28]

We are dependent upon all these resource cycles. Our bodies are designed to be integrated into them. Today, we are altering these cycles on a global scale. Predicting the consequences of these changes is difficult. Wisdom and prudence would suggest that we use caution in changing these cycles since we know we depend on them for life, and we cannot predict the effects of the changes we are making.

Energy Flow and Exchange

The sun is the principal source of energy for all life on earth.[29] It provides a steady, daily input of usable energy. As we noted, photosynthetic organisms (plants, algae, and some bacteria) are able to capture solar energy and convert it into chemical energy in the form of organic molecules (sugar, starch, protein, etc.).[30] These organisms are called autotrophs or primary producers, and they form the energetic foundation of all ecosystems. They are consumed by animals, called primary consumers (herbivores), who then utilize their organic molecules as their energy source for maintenance, growth, and reproduction. These animals may, in turn, be consumed by still other animals called secondary

---

28. Krebs, *Ecology*, 570–71.

29. Nuclear and geothermal energy contribute a small amount to the overall energy budget of the planet. Otherwise all energy utilized by humans and all living things on earth ultimately comes from the sun. Even fossil fuels originally consisted of plant material formed through photosynthesis using the sun's energy. The plant material then decayed and, over millions of years under conditions of heat and pressure within the earth's crust, was converted to coal, oil, or natural gas.

30. Plants and other photosynthetic organisms capture only about 1 percent of the solar radiation that reaches the surface of the earth (Odum and Barrett, *Fundamentals*, 109). The remaining energy is not "wasted," however, since it goes into driving the water cycle and maintaining surface, ocean, and atmospheric temperatures at appropriate levels.

Ecology for Dusty Earthlings

consumers (predators or carnivores).³¹ Top predators may prey on secondary consumers, forming a third or even a fourth trophic level.

An example of this is a coastal ocean ecosystem. Here sunlight is captured by phytoplankton—microscopic plants and bacteria in the water—and converted into biological molecules (primary producers). The phytoplankton is eaten by zooplankton—microscopic animals also drifting in the water. These are in turn eaten by small fish such as anchovies and sardines (secondary consumers). The small fish are in turn eaten by larger predatory fish such as tuna, shark, and marlin, which are at the top of the energy pyramid (tertiary consumers).

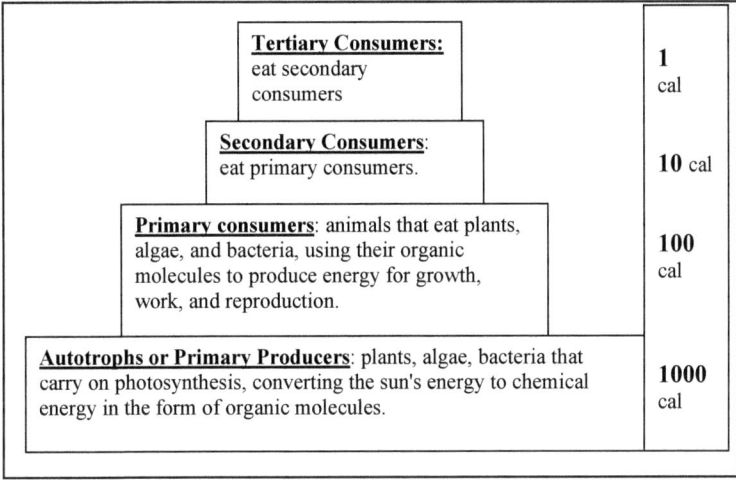

Figure 1: The Energy Pyramid

These levels form what is called the *energy pyramid*, which shows how energy flows through an ecosystem.³² (See Figure 1.) The laws of physics say that energy transfers between these levels are never perfectly (100 percent) efficient. In fact, on average, in nature, these energy transfers are only around 10 percent efficient. That is, only about 10 percent of the energy consumed from one level is actually utilized by the consumer at the next level for maintenance, growth, or reproduction.

---

31. It should be noted here that many organisms operate at more than one level. Humans, for example, are omnivorous. We eat plant material (primary consumers) or meat (secondary consumers). Many other organisms operate at more than one trophic level such as foxes, bears, and chimpanzees.

32. Odum and Barrett, *Fundamentals*, 102.

Dusty Earthlings

The remainder is dissipated into the environment in the form of heat or waste (feces, urine). This is indicated on the scale on the right side of Figure 1. One thousand calories stored by a primary producer such as phytoplankton, grass, or trees, once passed up the pyramid, yields only one calorie of energy for a tertiary consumer. Energy exchange as represented by the trophic pyramid appears to be a basic feature of all ecological systems.[33]

One can see that organisms operating at higher trophic levels place a larger energy load on the system than those operating at lower levels. In the case of humans, eating meat (secondary consumption) requires around ten times as much energy as eating at the primary consumer level (vegetables, fruits, grains, etc.). And eating the meat of secondary or tertiary consumers such as tuna, shark, and marlin, requires 100–1,000 times as much energy as eating at the primary level. Ecologically aware Christians who recognize their integration within the ecosphere and the higher pressure that consuming at higher trophic levels places on the ecosystem, should consider eating at lower levels. This is one reason (among others) that I have become a vegetarian. If I can consume at a lower level, it reduces the load I place on the ecosystem, making room for other humans and animals to live on the planet with me—an application of the principle of energy exchange and the commandment "Love your neighbor as you love yourself" (Lev 19:18; Mark 12:31).

Food Webs

All organisms exchange energy and nutrients with other living things within their ecosystems by eating, being eaten, producing waste, and dying. Networks of these exchanges make up what are called *food webs*. By understanding food webs, we can understand the multidimensional nature of energy and material exchange among organisms within ecosystems. All living things in all ecosystems on earth are embedded in food webs, including us humans. Even if you live in the middle of a city and never see another living thing except humans, pets, pigeons, and pests, you are still integrated into a food web. Simply put, it's the way things are, and all life depends on it.

Food webs are complex, but a basic understanding of them can help us see how ecosystems work, how we fit in, and how our behavior

33. Ibid., 103.

Ecology for Dusty Earthlings

affects them. A basic food web is one that occurs in the Southern Ocean around Antarctica. It is diagramed in Figure 2 below. Notice how the plankton, krill, and copepods, all of which float freely in the ocean, form the *base* of this food web. Phytoplankton carry on photosynthesis and so incorporate energy from the sun into their bodies (primary production). Zooplankton feed on phytoplankton. Copepods are microscopic animals that feed on all plankton and are in turn eaten by krill. Krill are small shrimp about an inch long that live in the Southern Ocean and can occur in huge swarms. Baleen whales feed on krill and on small fish. Toothed whales include killer whales and sperm whales and have no other predators except humans. With humans they are at the top of the food web and are called *top predators*. Notice the parallels between the food web and the energy pyramid (Figure 1). Energy and nutrients pass up the food web from the primary producers (phytoplankton) to secondary, tertiary, and higher consumer levels (fish, seals, whales, humans). In general, food webs can be thought of as having two parts: (1) the *grazing food web* consisting first of green plants, algae, and phytoplankton, then herbivores (plant eaters—zooplankton, copepods, and krill), and higher consumers; and (2) the *detritus food web*, which begins with waste and dead organic matter (detritus) from organisms at all levels, then organisms that feed on that detritus, and their predators.[34] These two components form the overall food web. Figure 2 shows only the grazing food web of the Southern Ocean. Material enters the detritus food web through the waste or death of animals at all levels. This waste and dead material sinks into the depths where it is broken down by bacteria and fungi and converted into nutrients. Upwelling currents bring these nutrients back to the surface where phytoplankton can use it, passing it back into the grazing food web, and so on.[35] Through these complex relationships and processes, resources are recycled again and

---

34. Ibid., 108.

35. Upwelling currents that bring nutrients up from the depths are important in oceanic ecosystems and are responsible for a great deal of productivity. For example, the California Current off the coast of California is a cold current. As it moves south into warmer regions, it tends to sink, displacing deeper water that then moves upward. This upwelling of colder water brings nutrients to the surface, increasing phytoplankton, and the smaller fish that feed on it. These in turn supply larger commercially valuable fish. As a result of this happy circumstance, fisheries off the coast of California and Baja California have been productive food sources for humans.

again, all powered by energy from the sun, allowing the ecosystem to sustain itself over time.

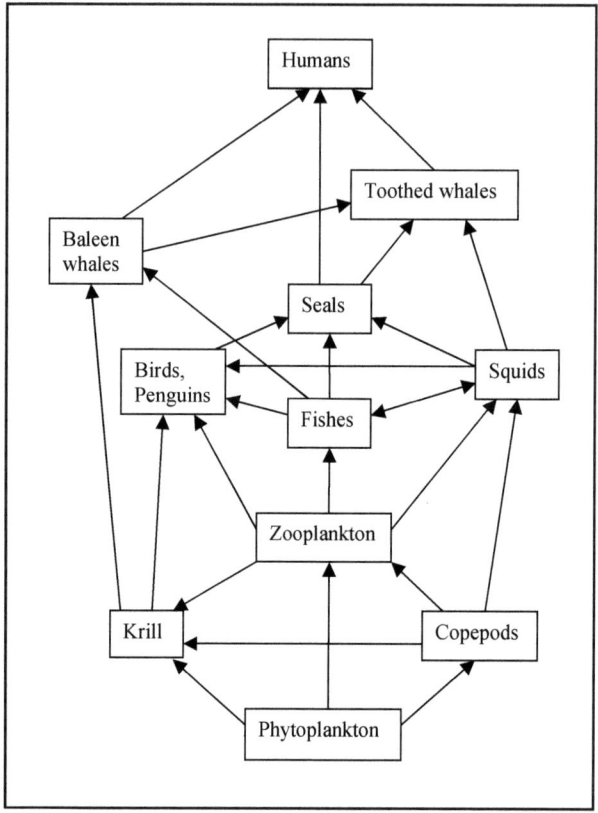

Figure 2. A simplified Southern Ocean food web (after Neil A. Campbell and Jane B. Reece, *Biology*, 7th ed. [2005], p. 1163. Used by permission of Pearson Education, Inc., Upper Saddle River, NJ). Arrows show how nutrients and energy pass through the web. See text for details.

We humans function as herbivores or carnivores in the grazing food webs of every ecosystem we inhabit. Compared to other animals, this is one of our great advantages. We are not confined to one specialized food source or place in the web; we are omnivores, meaning that we can consume both plants and animals and even fungi (mushrooms and yeast). Thus, we are less vulnerable than other species to scarcity and perturbations in the ecosystem. For example, in the Southern Ocean, humans are top predators, feeding on large and small fish, blue whales, killer whales, and seals. Currently, however, due to overexploitation, whale and seal hunting has been curtailed. Now humans have moved down the food web and are harvesting krill.[36] Krill are used for aquarium food, fish bait, food for fish farming, and krill oil is now marketed as a health food containing omega-3 fatty acids and other allegedly healthful substances. The human krill harvest appears to be sustainable at present, but if it increases substantially, it could impact the food web by depleting the krill that form part of its base. If this were to happen, it would ripple through the entire food web, impacting the populations of all animals involved.

For us humans, there are at least four aspects of food webs, energy pyramids, and the nutrient cycles that are important. First, the sustainability of ecosystems depends on the timely passage of nutrients and resources through these systems. In this way, ecosystems can persist long term. Unfortunately, current human industrial systems of material production and consumption are not cyclic but linear—what might be called an "extract-manufacture-sell-use-discard" regime. In some cases the rate at which discarded waste is produced exceeds the absorptive capacity of the ecosystem (CO_2, for example). In other cases the form of the discarded waste—various plastics, for example—are only slowly degraded by the system.[37] This material cannot enter the detritus food web, or it enters much too slowly to be recycled in a timely fashion, or it consists of toxic substances that injure or kill organisms involved in the food web.[38] Thus, human activity has resulted in the accumulation of large amounts of material in the form of trash and garbage in landfills and in the general environment that cannot be cycled through the food web or does so very slowly. Our industrial system operates in such a way that "all the energy and much of the matter that passes through the human industrial system is permanently dissipated into 'the environment' never to be used again,"[39] or at least not for a very long time. We

---

36. Krebs, *Ecology*, 463.

37. Seabirds such as albatrosses who feed on small surface fish at sea eat plastic objects such as syringes and toys that float among the vast quantities of human trash circulating on the oceans. The birds mistake the objects for food. They also may feed the objects to their young, resulting in death. This phenomenon poses a threat to these seabird populations (Kostigen, *You Are Here*, 155).

38. These toxic substances include elements such as mercury, cadmium, lead, arsenic, and compounds such as dioxin, polychlorobiphenols, and a multitude of other organic and inorganic compounds.

39. Wackernagel and Rees, *Ecological Footprint*, 44.

need to begin moving toward redesigning our industrial systems so that all materials are biodegradable and recyclable, either by us or by the detritus food web. This is a fundamental aspect of sustainability—being able to maintain ourselves and other creatures within the food web of the ecosystem over time without degrading it.

Actually, we may be in the early stages of making this transition. Recycling has become part of life for many of us. We currently recycle aluminum, paper, cardboard, some plastics and electronics, and so on. Some private, public, and commercial entities compost their food and yard waste. But at present this is inefficient because, except for metals and compostables, we are trying to recycle materials and products that are not designed for it. Consequently, their recycling requires a lot of energy and other resources.[40] In effect, recycling is an afterthought. But public awareness and demand is rising, manufacturers are becoming better informed, and research is ongoing. We the public can help by recycling everything we can and by demanding more recyclable materials. In this way, we should see more and more materials and products specifically designed for recycling and biodegradability in the future.

Second is the principle of *biological magnification*. This is where certain chemicals and heavy metals (for example, mercury and lead) accumulate and become more concentrated as they pass up the food web, especially from producers through primary and secondary consumers to top predators (humans). The classic example of this is DDT.[41] Due to its inexpensive manufacture, ease of use, and persistence in the environment, this pesticide was used extensively for crop dusting and insect control from the 1940s to the 1960s. It was consumed by organisms at several levels, and, due to its persistence, it circulated widely through the food web. As DDT passed up the food web, it accumulated in higher and higher concentrations in the tissues of animals at higher trophic levels. In other words, it became *magnified* or more concentrated (hence the name biological magnification). Levels of DDT in top predators such as eagles and some humans were found to be several times higher than in insects or microorganisms at lower levels.[42] It turned out

---

40. I should note here that contrary to a popular myth, for almost all materials, recycling requires less energy and resources than making things from scratch.

41. DDT stands for dichlorodiphenyltrichloroethane.

42. The presence of DDT in humans caused concern about possible toxic effects although none were ever identified at the levels observed. Thankfully, DDT was banned

that DDT made the shells of birds' eggs fragile, resulting in breakage in the nest and lower reproductive rates. The bald eagle, for example, was brought nearly to extinction due to this problem. In addition, DDT was found in significant concentrations in animals thousands of miles from where it was being used. This meant that DDT was being spread far and wide by migratory animals, rivers, winds, and ocean currents and was getting into places where we never intended it to be. At the same time, significant resistance to DDT was developing in the insect species it was meant to control, such as crop pests and the Anopheles mosquitoes that transmit malaria. As a result, DDT's effectiveness was declining. Scientists recognized the damage DDT was doing to bird populations and the potential threat it posed to other ecosystems and animals including humans.[43] Finally, in 1975, DDT was banned in the United States and Europe, and its use has dropped considerably around the world. Thanks to this ban, today bald eagles, golden eagles, and other threatened bird species are recovering. Other toxins, such as the heavy metals cadmium, mercury, and lead, also undergo biological magnification. We should keep this phenomenon in mind, especially in light of the thousands of new chemical compounds that we synthesize each year and release into ecosystems, particularly in pharmaceuticals and electronic waste.[44]

Third, overharvesting of food at one level in a food web can have consequences throughout the system. An example of this is a case of overfishing along the southern coast of Alaska and the Aleutian Islands in the northern Pacific. In the 1990s sea otters living along these coasts declined rapidly. The reasons for this are complex, but ecologists observed that at the same time, seals and sea lions also declined, probably due in part to humans overexploiting fish species that are food sources for them. Seals and sea lions are a favorite prey of killer whales, so when the seals and sea lions declined, the whales turned to sea otters for food, reducing sea otter populations. Sea otters feed on sea urchins, which in turn feed on the kelp (seaweed) forests that occur along these coasts.[45]

---

and human levels began to fall before any ill effects occurred.

43. See Carson, *Silent Spring*, 178–81.

44. For a good overview of the DDT story, see McGinn, "Combating Malaria," 536–62.

45. In these ecosystems sea otters are thought to be a "keystone" species. This means that their role in the ecosystem is larger than their relative abundance would

The reduction of the sea otters resulted in an increase in sea urchins, which reduced some kelp forests that, in turn, serve as nurseries for the juvenile forms of some commercially valuable fish.[46] Thus, the effects of overfishing rippled through the food web, leading to reduced productivity of the system for the fishing industry, which was already overtaxing the system. The take-home lesson is that the harvesting of wild species, whether it is fish, krill, or whatever, must be done thoughtfully, using the best ecological knowledge we have. Humility and prudence are in order. Such careful management and self-restraint can help maintain natural populations as well as resources for commercial harvesters, sportspeople, and for the creatures themselves.

Fourth, although ecologists are learning more all the time, our knowledge of food webs is limited. Removing species, adding nonnative species, or changing the food web in other ways often has unintended consequences that are difficult to predict. For example, Lake Victoria in East Africa harbored the greatest diversity of freshwater fish species in the world—at least four hundred different types, many of them found nowhere else in the world. In 1954, the Nile perch was intentionally introduced as a potential food source for native peoples who live around the lake. An aggressive top predator, the Nile perch can grow up to six feet long and weigh over two hundred pounds. They have proliferated and now dominate the lake. Almost all of the four hundred native fish species have either disappeared or are rarely seen. Many are believed to have become extinct. In addition, the oxygen in the lake has declined to very low levels. Some ecologists have called this "the greatest devastation ever wrought by an introduced predator."[47] The large Nile perch population has spawned an active commercial and sport fishing industry around the lake, which has generated jobs and some wealth but has also caused significant social disruptions in human communities. The perch is now being overfished and is in decline. As a result a few native species are making a comeback, but the perch's decline is jeopardizing the future of the fishing industry that built up around it, potentially

---

suggest. Their presence is necessary to the integrity of the ecosystem because they play a "key" role in the food web. Identifying keystone species is difficult. Their role in an ecosystem may be complex, involve more than one species, be context dependent, and vary over time. Thus, their identification by field ecologists can be challenging. See Powers et al., "Challenges," 609–20.

46. Krebs, *Ecology*, 471–72; Estes et al., "Complex Trophic Interactions," 629–32.

47. Molles, *Ecology*, 403–4.

causing further social and economic disruption. The future of the Nile perch in Lake Victoria is uncertain.[48] Again, prudence, humility, and care are in order as we humans manipulate and exploit food webs.

Biodiversity and Ecosystem Stability

The concept of *biodiversity* is complex, but for my purposes here, I will use the word simply to denote the number and variety of species present in a given ecosystem. Humans are causing species loss and reducing the biodiversity of ecosystems by various means: habitat destruction, overharvesting, introduction of nonnative species, pollution, and modification of abiotic factors such as temperature, salinity, acidity, the nitrogen cycle, the carbon cycle, and the water cycle.[49] In order to prevent loss of biodiversity, ecologists want to know how species diversity in ecosystems is sustained and what makes ecosystems stable (resistant to disturbances) and resilient (able to recover when disturbances do occur). If ecologists can answer these questions, then perhaps they can offer us better ways to manage ecosystems and to protect the wondrous diversity of species that God has created and placed in our care. A lot of research has been done on this, and as a result, it is generally accepted that higher levels of biodiversity (a greater number and variety of species) is associated with increased ecosystem stability and resilience.

Because they are associated with reduced biodiversity, human-made ecosystems such as agrarian monocultures (large areas of single crop farmland) and urban ecosystems (cities) may, in the long run, be less stable and resilient.[50] In fact, from an ecological perspective, we can see that these human-constructed ecosystems require continuous high inputs of energy (fossil fuel, electricity), materials (machines, chemicals, clothing, consumer goods, building materials), and nutrients (water, fertilizers, food) in order to maintain their precarious stability. If such input is cut off or reduced, these systems collapse or degrade quickly. One can imagine what would happen to, say, Los Angeles if all input of water or fossil fuel was cut off. The city would rapidly become chaotic and disintegrate. In this way, these human-made ecosystems

48. The story of the Nile perch in Lake Victoria was dramatized in the film *Darwin's Nightmare* produced by Hubert Sauper in 2004.
49. Hooper et al., "Effects of Biodiversity," 3–4.
50. Ibid., 22.

Dusty Earthlings

are vulnerable to disturbances and may not be stable in the long term. We need a lot of creative thinking and innovation about how to make human-made urban and agrarian ecosystems more biologically diverse and hence more stable and resilient.[51]

Ecosystem Services

Ecosystem services (ES) are "properties of an ecosystem that either directly or indirectly benefit human endeavors."[52] Of course, ecosystems provide services to all living things, but we humans want to know what ecosystems do for us. Ecosystem services include such benefits as climate regulation, water purification, waste treatment, soil formation, erosion control, nutrient recycling, pollination, recreation, and so on.[53] The importance of ES was spotlighted in a 1997 paper when a group of researchers estimated the global economic value of ES to humanity at about 33 trillion dollars per year (range 15–54 trillion), an amount comparable to the global GDP of 18 trillion at the time.[54] In other words, ecosystems provide a lot of benefits to us "free of charge." *The Economics of Ecosystems and Biodiversity* (TEEB) report in 2010, commissioned by several organizations, addressed the relative values of ES and recommended that they be integrated into economic reporting and decision-making by businesses, governments, and other agencies.[55] Currently, our economic system ignores these services; yet it is completely dependent on them. Economists call them *externalities* because they see them as *external* to the markets that are the heart of the economy. I am not suggesting that all these services be converted into markets, although perhaps some ought to be. The point is that we are ignoring them while being wholly dependent on them. We are using God's resources without counting the cost (Luke 14:28). Perhaps understanding and taking into account the economic value of creation's services to us will help us appreciate them, encourage gratitude and sharing, and help us do a better job of managing them.

---

51. There is good work going on in this area. See, for example, Gottlieb, *Reinventing Los Angeles*.
52. Hooper et al., "Effects of Biodiversity," 7.
53. Costanza et al., "Value of the World's Ecosystem," 254
54. Ibid., 259.
55. Sukdhev et. al.,*TEEB (2010)*, 28; see also Stern, *Stern Review*.

Population Dynamics, World Human Population, and Carrying Capacity

By population dynamics, I mean how biological populations grow, contract, or remain stable, how they are distributed, and how habitat, consumption, waste, and other factors interact to affect population size and distribution. Ecologists have done a lot of work on animal populations and have developed some basic principles about their dynamics. Since we humans are animals, too, we are subject to these same principles. Here I want to discuss a few of these principles and some of their implications for how we might live sustainably long term within the limits of the ecosphere.

A word of caution is in order here. Animal population dynamics are complex, but those of humans are even more complex. Animals are affected by factors such as predator-prey relationships, disease, parasites, habitat availability, mate selection, climate variation, nutrition, and stochastic factors (random events and disturbances). Humans are affected by all of these plus culture, technology, militarism, nationalism, economics, politics, religious beliefs, social structures, and so on.[56] Also, humans make choices—individually and corporately. We cannot predict what those choices will be—what wars we will fight, what economic ups and downs might occur, what cultural trends and shifts might develop, what kind of men women will prefer, how many children they will want to have, and so on. As a result, these principles of population dynamics have only partial explanatory value and little predictive value. We can say a fair amount about how population dynamics work, but we can make only tentative future predictions based on that.

Overall, however, there is one "law" of population dynamics that we can rely on. It is stated succinctly by Harold Dorn: "No species has ever been able to multiply without limit. There are two biological checks upon a rapid increase in number—a high mortality and a low fertility. Unlike other biological organisms, [humans] can choose which of these checks shall be applied, but one of them must be."[57] This is a stark statement, but it is true. We humans are complicated, but as eco-physical beings we still have basic needs for food, space, shelter, and resources that only the ecosystem can provide. Like all earthly crea-

---

56. Cohen, *How Many People*, 123.
57. Dorn, "World Population Growth," 290.

tures, we cannot increase without limit. A corollary to this "law" is that human populations cannot increase in wealth, material consumption, waste output, travel, luxury, and technology without limit. Again, not only our numbers but also our standard of living is wholly supported by the ecosystem, and the ecosystem has limits. Sooner or later, in one way or another, rising numbers and increasing affluence will encounter limits. Unlike other organisms, we humans can choose if, when, and how these limits will be imposed—whether by voluntary self-control and sharing among ourselves or by government fiat, famine, disease, privation, war, or death.[58]

Most all animals and plants have high potential for reproduction. If unchecked, they could rapidly fill the earth with their own kind. A maple tree can produce thousands of seeds each year. If all these seeds germinated and grew into adult trees that in turn gave rise to thousands more, the earth would soon be taken over by maple trees. But this does not happen; because of limits on maple tree growth—limits on space, nutrients, and light, as well as factors such as competition and predation—most of the seeds die and never germinate. Of those that do, most of the seedlings do not survive. Only a tiny percentage of the seeds eventually grow into adult trees that reproduce. The fecundity of life is checked by the limits of the created order.[59] Similarly, all animals are subject to the same limiting effects of the ecosystem. We humans likewise have a high capacity for population growth, but similar limits impose on our population growth—and on the growth of our wealth and living standards.

A common model for the behavior of animal populations is what is called the *logistical model* (see Figure 3.) This model assumes the existence in an ecosystem of what is called the *carrying capacity*, which is defined as the "maximum population of a given species that can be supported indefinitely in a specified habitat without permanently

---

58. When animal populations do approach ecological limits, those individuals and subpopulations that live in poorer habitat and are marginally supported are the ones who suffer and are eliminated first. Dominant populations that occupy the better habitat and control more resources are able to survive longer. Hence, as human populations encounter limits, it will be the poor, the powerless, the weak, and the marginalized of the world who will suffer first. We the wealthy (for now) will be able to survive longer.

59. Glacken, *Traces*, 625.

Ecology for Dusty Earthlings

impairing the productivity of that habitat."[60] That is, a given habitat has a limited amount of resources such as food, water, space, places for shelter (nests, burrows, etc.), and can absorb waste up to a certain limit for the particular species in question.[61] Also, parasitism, disease, predators, prey, available mates, hiding places, immigration, emigration, and disturbances (storms, fires, floods, epidemics) can limit the number of individuals a habitat can support over time. In reality, carrying capacity constantly changes, shifting up or down as it interacts with living populations and with other biotic and abiotic factors. Another assumption is that all animal species have a finite maximum reproductive rate. This is, of course, true, but actual rates almost always vary at levels below the maximum rate.

Under controlled conditions (such as a laboratory), an animal population introduced into a suitable habitat will usually (but not always) follow something like logistical growth. After a lag period, the population increases slowly at first, then more rapidly until a steep, rapid rise occurs. As the population approaches carrying capacity, its growth slows and then levels off as various limiting factors come into play. Finally, it stabilizes at or below carrying capacity.[62]

---

60. Wackernagel and Reese, *Ecological Footprint*, 49.

61. Molles, *Ecology*, 259.

62. The logistical growth curve makes several other assumptions: (1) "each new individual has an instantaneous effect on the population's growth rate, an assumption that is almost always false"; (2) individuals are added to the population at a constant rate at any given population level, an assumption that is often not the case; and (3) the carrying capacity is constant, which, as noted in the text, is not true (Van Dyke, *Conservation Biology*, 215).

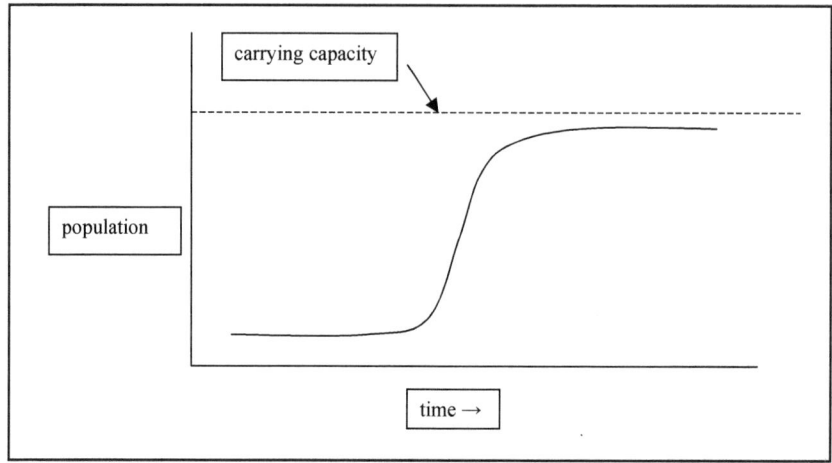

Figure 3. The logistical model of population growth

In real ecosystems, animal populations usually do not follow the logistical model. If a rapid growth phase occurs, the population may not level off but may follow several different patterns (see Figure 4.) The environment may change. There may be variations in temperature or rainfall, fires, floods, or disease, or the population itself may interact with the ecosystem in such a way as to change the carrying capacity through overexploitation of food supplies, territoriality, changes in reproductive patterns, fighting, increased predation, disease, parasitism, or random chance. Or the population may overshoot the carrying capacity, leading to resource shortages, overcrowding, accumulation of waste, or negative feedbacks that reduce the population. The population may crash; or it may enter a period of fluctuation in which positive and negative feedbacks and lag times cause large ups and downs (something like the lynx and snowshoe hares); or the population may enter a decline leading to a steady or variable state at a lower level, or even to extinction; or the population may vary in an apparently random fashion. There are myriad variations.

Predicting the future of the global human population is difficult, if not impossible.[63] Intrinsic factors such as decreased birth rate (as women become better educated and empowered), increased crowding in cities, cultural and social changes, technology, and war, or extrinsic factors such as disease (AIDS, for instance), storms, earthquakes, volcanic

---

63. Krebs, *Ecology*, 143; Cohen, *How Many People*, 109–10.

eruptions, climate change, or limitation of food and energy supplies, could slow growth or cause decline. Conversely, new technology, increased production of food and consumer goods by modern industry, reduced mortality (increased longevity), and higher reproductive rates could lead to higher populations.[64] We humans have shown that we have enormous power to use technology to modify the ecosystem, extract resources from it, and transfer them from one place to another.[65] Chemical and genetic manipulation of domestic animals and plants en masse has led to the production of enormous amounts of food to the point where large segments of the world population today are overfed (while others are underfed or starve).

Remarkably, despite its complexities, the world human population seems to be following the logistical model, at least so far. After a long lag period, we entered a rapid growth phase beginning around 1800. The world human population growth rate peaked at slightly above 2 percent per year in the 1960s. Since then the rate has declined. In 1990 it was 1.6 percent per year,[66] and it continues to decline slowly, placing us at present in the part of the curve that is beginning to level off. But we do not know what the rest of the curve will look like (Figure 4). The population could continue to go up, level off, go down, fluctuate, or even crash.[67] The United Nations estimates that the world's human population will increase from its current figure of 7 billion to between 7.8 and 12.5 billion and stabilize there sometime around the middle of this century.[68]

---

64. Cohen, "Population Growth," 356.

65. At present this is almost totally dependent on fossil fuels, the burning of which produces CO_2, the principal greenhouse gas that is driving global climate change. If this proves to be the serious problem many think it is, whether we will be able to change to another less problematic energy source to support our massive global transport system on which we depend remains to be seen.

66. Cohen, "Population Growth," 349.

67. It is unlikely that the population would stabilize at a certain level and remain there. This would require that births and deaths remain equal over time, which is highly improbable.

68. Cohen, *How Many People*, 368.

Dusty Earthlings

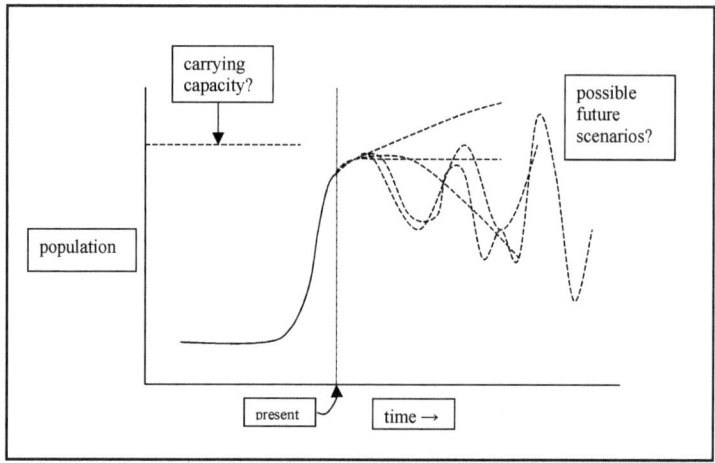

Figure 4. World human population. The solid line represents population growth to the present. The dotted lines represent possible future population behavior.

We do not know what is the maximum number of people the earth can support (carrying capacity) nor where we are in relation to it. Estimates of human carrying capacity have varied from one billion to over a trillion, but more rigorous estimates cluster between four and sixteen billion.[69] If this range turns out to be more or less accurate, then at our current seven billion we may be approaching carrying capacity. Ecologist Charles Krebs thinks that we are currently at or near it.[70] The authors of the book *Limits to Growth* believe we are currently in "overshoot"—beyond carrying capacity—and if current trends continue, we are headed for a decline or perhaps a crash.[71] Using their ecological footprint model, which I will explain shortly, Mathis Wackernagel and William Reese also conclude that we are in overshoot.[72] Demographer Joel Cohen, despite his caution in making any predictions at all about human population, warns that we may now be "entering a zone where limits on the human carrying capacity of Earth have been anticipated and may be encountered."[73]

69. Cohen, "Population Growth," 351.
70. Krebs, *Ecology*, 590.
71. Meadows, Randers, and Meadows, *Limits of Growth*, xiv.
72. Wackernagel and Rees, *Ecological Footprint*, 90.
73. Cohen, "Population Growth," 351.

Ecology for Dusty Earthlings

Whatever the future holds, we should accept the "law" that for earth creatures like ourselves, numbers and affluence cannot increase without limit.[74] As we have noted, we humans are more complicated, and many factors are at play.[75] But these factors cut both ways. They can increase carrying capacity or decrease it; they can accelerate population growth or slow it down. Despite these complications, we remain eco-physical beings embedded within the ecosphere, and as such we cannot escape the limits of earthly life. Our technological and economic success and the opulence and wealth that *some* of us enjoy *for the moment*, do not exempt us from ecological realities on a finite planet. As demographer M. A. Hixon notes, "Some human societies have behaved as if no limits exist."[76] Hixon is talking about affluent societies like ours here in America. Whatever we may say, our lifestyles and behavior clearly embody the belief that growth and affluence have no limits. But if what I am saying here is true (or even partly true), then this is a false belief. We are not limitless beings. Like all God's creatures, we are and always will be subject to the limits of growth, even if they are complex and changing, and even if we don't know exactly what they are, and even if our current population models and predictions are flawed. We cannot escape earth. We cannot avoid our eco-physical nature.

Because carrying capacity is difficult to determine, especially for humans, Mathis Wackernagel and William Rees have pioneered a new approach called *ecological footprint* analysis. Here, the ratio is inverted. Instead of trying to calculate how many people could live in a given area of land, they calculate the area of land required to support a given population of people who are living at a given standard of living.[77] This allows them to take account of resource transport, manufacturing, waste absorption, living space, recreation, travel, and so on. Using their current methods, Wackernagel's group, the Global Footprint Network, estimates that worldwide, the average per capita footprint is about 2.7 hectares (6.8 acres) per person.[78] The total amount of ecologically productive land and sea available is about 11.2 billion hectares, or about

74. Hixon, "Carrying Capacity," 530.
75. Bratton, *Six Billion*, 38.
76. Hixon, "Carrying Capacity," 530.
77. Wackernagel and Rees, *Ecological Footprint*, 5, 9–13; Wackernagel and Kitzes, "Ecological Footprint," 1031.
78. Global Footprint Network.

1.8 hectares (4.5 acres) per person.[79] Thus, by these calculations, the global human population is now living about 50 percent beyond the capacity of the planet to support it long term. Or put another way, we are using about 1.5 planets' worth of resources. The average footprint for us North Americans is about 8 hectares (20 acres) per person.[80] This means we are taking much more of the earth's resources than others.

Ecological footprint analysis is in its infancy, and its calculations are at best approximate. The Global Footprint Network and others will continue to improve their methods. So far the results of footprint analysis are not favorable. Although they are approximate, we should not ignore them. Counting the cost by seeking to determine human demands on the ecosphere and its capacity to meet those demands is essential if we are to be wise, faithful, and just caretakers of God's earthly creation.

Interestingly, the problem of carrying capacity appears in the Bible. In Genesis 13, Abram and Lot encounter the ecological limits of a part of Palestine. When they went up from Egypt into Palestine, they found that "the land could not support them while they stayed together, for their possessions were so great that they were not able to stay together. And quarreling arose between Abram's herdsmen and the herdsmen of Lot. The Canaanites and Perizzites were also living in the land at that time" (Gen 13:6–7). This is a clear-cut case of overshoot of the local carrying capacity. The ecosystem was limited in size, resources, and productivity and could not support the pastoral lifestyles of Abram, Lot, their entourages, and the other tribal groups living in the area. "In one of the oldest recorded human encounters with environmental carrying capacity, responsible [Abram] gives Lot the option of taking the best grazing territory . . . Lot chooses the rich river valley, while Abram turns west to the land God has selected for him in Canaan."[81] In this case, there was open land available outside of the local region to relieve the overload. But in today's world, there is almost no open land left. We do not have the option of splitting up and moving to unoccupied territory as Abram and Lot did. We live in a more or less full world and have to learn to live together peaceably and distribute resources equitably within that world—or, perhaps, face something like the quarreling that arose between the herdsmen of Abram and Lot. Today, given our highly

---

79. Wackernagel and Kitzes, "Ecological Footprint," 1033.
80. Global Footprint Network.
81. Bratton, *Six Billion*, 45.

Ecology for Dusty Earthlings

developed weapons technology, such quarreling could be deadly. But what is significant is that in this story, the Bible recognizes the reality of ecological limits.

The lesson that we affluent Western Christians should take home from this is that, given our high levels of consumption and waste, we are impacting the world's carrying capacity in a way that is unjust. We are taking more than our share of the earth's resources for ourselves at the expense of other people and to the detriment of God's creatures and ecosystems.[82] When this is considered in light of the teachings and example of Jesus Christ, it seems clear that we rich Christians have some soul searching, repenting, and changing to do. More on this in chapter 7.

The Second Law of Thermodynamics or the Law of Entropy

Physics contains what is called the *second law of thermodynamics* or the *law of entropy*. This was developed in the nineteenth and early twentieth centuries by physicists studying engines and other physical systems that use energy to do work. Since I am arguing that we humans and all living organisms are physical creatures (or physical systems), it follows that physical laws such as the law of entropy apply to us. The word *entropy* means disorder.[83] Disorder means exactly that. A neat stack of books on a table has low entropy, a low level of disorder. If the stack is pushed over so the books scatter on the floor, its level of disorder increases. It becomes a less ordered, or a more disordered, system. The entropy of the books has increased. Probably anyone who has had a teenage son or daughter and has seen their bedrooms knows what entropy looks like. The second law of thermodynamics, the law of entropy, says that in a closed system,[84] with any change that occurs, *overall entropy (disorder) always increases*. Actually we already know this intuitively through our experience of the world. Examples abound. Heat flows from a warmer body to a colder body. Water in a glass at room temperature does not suddenly freeze into ice or start to boil; it warms or cools until its

---

82. Bauckham, *Bible and Ecology*, 74.

83. Actually, the concept of entropy is more complex, but it is legitimate to think of it as disorder.

84. By "closed system" here, I mean a system that is isolated from its surroundings with respect to energy, matter, and information. That is, it does not exchange any of these things with its surroundings.

temperature is the same as its surroundings. If I drop an egg on the floor and it breaks, it will never reassemble itself back into an intact egg, no matter how long I stand there and wait for it. Since overall entropy must always increase in a closed system, a decrease in entropy in one place means that entropy must increase somewhere else such that overall net entropy increases. Usable energy input is always required to cause any decrease in entropy,[85] and usable energy is always required to maintain that decrease in entropy in relation to the surroundings. For example, your refrigerator maintains a lower state of entropy by keeping the temperature colder inside than outside. It requires an input of electricity (energy) in order to produce this state of affairs and a continuous input to maintain it. In the case of the stack of books I noted above, the local focus of decreased disorder (entropy) represented by the stacked books occurred as a result of the energy used by the person who stacked them. When the books were pushed off the table, that energy was dissipated into the room in the form of heat (and perhaps damage to the books or the floor) and represents an increase in entropy such that the overall entropy of the whole system of books, table, the persons who stacked and pushed them, the room, and the universe increases. The law of entropy has generally stood the test of time, and today, it stands as a "law" of physical reality, including eco-physical reality. No known technology allows us to transcend it. All human and natural systems must obey this law—no exceptions.[86]

As noted, energy input is required in order to maintain a local ordered system of low entropy in relation to its surroundings. The earth itself, when compared to the other planets and the universe as we know it, is a highly ordered, localized system with low entropy. The atmosphere, oceans, climate system, and biosphere constitute a highly ordered ecosphere. How can this be? It is because the earth receives a constant input of usable energy from the sun.[87] In this sense, the earth is not a closed system; it is an open system. (With respect to matter, the earth *is* a closed system. It exchanges no significant amount with

---

85. Usable energy is any energy that the system in question can use. For plants, this is sunlight. For us humans and other animals, this is the food we eat. Heat, for example, is generally not a usable form of energy by which plants and animals can maintain their low-entropy states.

86. Odum and Barrett, *Fundamentals*, 78.

87. The average amount of solar energy that reaches the earth's surface each day is about 288 watts per square meter (Houghton, *Global Warming*, 19).

its surroundings.) This solar energy is absorbed, reflected, and scattered by the atmosphere, the land, the oceans, and by living things such that a complex system of energy exchange exists in the ecosphere that keeps the earth stable and livable. Ultimately, the energy is dissipated back out into space (in the form of lower "quality" unusable energy such as heat) so that the earth's net energy content remains more or less constant.[88] Without the sun's input of usable energy, the earth would quickly fall into disorder; heat would dissipate into space, the oceans would freeze, the atmosphere would condense and freeze, and, except for an occasional volcano, the earth would become a frozen, lifeless planet. The earth is the wondrous, productive place that it is because of the sun's continuous energy input. We have complex weather, ocean currents, millions of species, grasslands, forests, streams, lakes, girls, boys, symphonies, computers, poetry, prayers, and baptisms all because the sun's input of energy allows the constant removal of entropy and the maintenance of order in the earth's ecosphere.

Living things also constitute highly ordered systems or localized centers of low entropy that maintain themselves by pumping entropy out into the world at large.[89] Again, entropy always tends to increase, even inside our bodies. Our bodies maintain an extremely high degree of internal order (low entropy) by constantly consuming food energy and using that food energy to "pump out" disorder (entropy) into the environment, mostly in the form of heat, but also as waste (carbon dioxide, feces, and urine). Nevertheless, the law of entropy slowly imposes itself, and bodily disorder gradually increases over time, as those of us who are aging can affirm. Similarly, an ecosystem can be viewed as an ordered system that requires energy input (sunlight) to "pump out" disorder to maintain itself.[90] It is important to remember that ecosystems

---

88. Interference with this system of energy exchange due to the buildup of carbon dioxide and other greenhouse gases in the atmosphere constitutes the problem of global climate change.

89. Belovsky et al., "Ten Suggestions," 348; Schneider and Kay, "Life," 447.

90. There are ecosystems that do not depend on sunlight as their source of usable energy. Hydrothermal vents in the deep ocean floor release water heated by hot rock and magma deep in the earth. The water is also rich in minerals. These vents, called "black smokers," support flourishing living communities. Chemosynthetic (not photosynthetic) bacteria utilize the minerals in the hot water to obtain energy and thereby provide the energetic and material base of these black smoker ecosystems. Hundreds of new species have been found in these systems.

and all living things (including us) are open systems. This means they continuously exchange energy and entropy with their surroundings.

An important corollary to the law of entropy for living systems such as our bodies (us) and for ecosystems is that they are *dynamic*—constantly changing over time. They are not static.[91] Energy input, energy dissipation (expenditure), and level of entropy change. Material exchange can also fluctuate over time. Each of us, for example, can gain or lose weight. Our health can improve or decline. We can contract a disease and recover from it. We can learn things and forget things. We can die. All these changes constitute us (you and I) as dynamic, low-entropy systems. In other words, as physical beings, we are what are called "far-from-equilibrium dynamic dissipative systems."[92] We are "far-from-equilibrium" in that we are localized foci of low entropy and highly organized matter compared to our surroundings. That is, we are *not* in equilibrium with our surroundings with respect to entropy, energy, information, or the concentrations and arrangements of matter.[93] We are dynamic in that we are constantly changing. And we are "dissipative" in that we must constantly take in and dissipate energy to maintain ourselves in our low-entropy state. Ecosystems are similar. They do not remain static. They can flourish or decline, grow or contract, and like us, they can die.

Modern human-constructed systems and artifacts such as cities, corporations, farms, families, automobiles, churches, universities, nations, and computers are highly ordered systems that must obey the second law of thermodynamics. These systems and the humans who build and use them must consume usable energy, such as fossil fuel or electricity or the energy of human activity, and dissipate it into the environment in the form of unusable energy (heat, CO_2, H_2O, waste) in order to maintain their order and function. But still they inevitably undergo decay and wear out, just as our bodies do. The ancient city of Rome at its height was the largest city in the Mediterranean region. To maintain itself, Rome expended prodigious quantities of energy, fighting many wars and importing fuel, food, building materials, clothing,

---

91. Belovsky et al., "Ten Suggestions," 348.
92. Schneider and Kay, "Life," 447.
93. This also applies to information. Our brains are stupendously concentrated foci of information compared to our surroundings.

water, information, and other resources from its surrounding empire.[94] Despite these efforts, ancient Rome did not endure forever but went into decline beginning in the fifth century AD. Today numerous cities much larger than ancient Rome require far greater inputs of energy and resources to maintain themselves. As historian J. Donald Hughes notes, "Cities import water and energy over hundreds, and food over thousands of kilometers. Forests are felled because cities need fuel, paper, and timber. No wilderness is so isolated as not to feel the influence of cities, from acids in the air and pollutants in the water to the noise of jet planes. City folk no longer depend only on local or regional resources; they are involved with the ecosystems of the Earth."[95] Today our cities are numerous and very large. They depend on vast inputs from around the globe. Fossil fuels are currently the principal source of energy we use to maintain low entropy in our cities and in all human social and economic systems.[96]

We, the minority of humans who live at higher standards of living (lower entropy levels), should understand that the maintenance of our affluence requires higher levels of energy and material input. Furthermore, since overall entropy must increase, our low entropy must produce higher entropy (disorder) elsewhere. For example, the 2.7 million people of San Diego County, California, where I live, are generally affluent, having low-entropy (highly ordered) lifestyles that require large inputs of not only energy and materials but also water. Since our local climate is arid, we import 90 percent of our water from elsewhere. Water drawn from the Colorado River has resulted in its severe degradation. The entropy of the Colorado River has increased in order to support the reduction in entropy we enjoy here in San Diego.

In the United States, the elevated (low-entropy) lifestyles to which we have become accustomed require that we "pump out" disorder to *somewhere else*. This somewhere else consists of the oceans, soils, rivers, and lands of our own nation but also of poorer, less powerful nations abroad. As Susan Bratton notes, back in the eighteenth and nineteenth centuries, England's population growth and industrial development were largely dependent on drawing resources from her colonial empire, which "fed the mother country with raw materials from all parts of the

---

94. Solomon, *Water*, 83–86.
95. Hughes, *Environmental History*, 218.
96. Laszlo, *Systems View*, 30.

globe and provided markets for her ever more wealthy manufacturers."⁹⁷ We could say that England pumped entropy out to her colonial empire in order to decrease her entropy at home. Our American war of independence was, in part, motivated by our dissatisfaction with these arrangements. But today, we Americans are doing the same thing, not by politically dominating weaker nations, as England once did, but by economically dominating them. Bratton warns further that the developing countries of today's world cannot support an increasing population or solve a food crisis the way England did in the past or the way we are doing now by effectively exploiting weaker people. Economically dominating weaker peoples and their lands in order to maintain our levels of consumption, material excess, and luxurious lifestyles, as we are doing, is not an option for nations like Congo or Haiti.⁹⁸ Unlike us, they have no one to exploit. (But they do have animals, plants, and ecosystems to exploit.) How this will play out, we do not know.

We should remember, however, that the earth is an open system with respect to energy. We receive a large amount of usable energy every day from the sun. This constant and reliable influx of energy is received by everyone everywhere. Thus, at least with respect to energy, poorer nations have the potential to improve their lives (lower their entropy) by utilizing solar energy (and its products—winds and tides), and developed nations, like the United States, have the potential to reduce their reliance on other nations for energy (fossil fuels) by doing the same.⁹⁹

---

97. Bratton, *Six Billion*, 31.
98. Ibid., 32.
99. We should remember that the exchange of energy and matter among ecosystems, cities, and human populations in today's globalized world is complex and involves not only physical and ecological principles, but also economic, technological, political, social, and cultural factors. For example, one could argue that a particular poor nation should be able to raise its standard of living on its own since it is receiving as much sunlight (or more, if it is in the tropics) as wealthy people do living in, say, North America. But it may not be able to do this because it lacks the means to purchase the technology (solar arrays), is embedded in a global economic system that channels resources away to wealthier populations, and/or because of political, social, and cultural factors.

The Principle of Limits

As Scripture and science tell us, there is a general truth about created existence: it is inherently limited. There is a great deal to be said about how this principle applies to our eco-physical lives, but I will focus on just two principles: "Shelford's law of tolerance" and human exploitation of wild animal populations.

Shelford's law of tolerance, named for its discoverer, Victor Shelford (1877–1968), says that "the distribution of a species will be controlled by that environmental factor for which the organism has the narrowest range of tolerance."[100] What this means is that if we examine the places where a particular species lives, we will find that it only lives in areas where its weakest and most vulnerable form or life stage can exist. For example, there is a fish called the arctic charr that, as its name implies, lives in the arctic in both freshwater and saltwater. The adult arctic charr can tolerate wide ranges of temperature, but their eggs cannot hatch in water above 20 degrees Celsius (65 degrees Fahrenheit).[101] Thus, even though the adults can tolerate warmer water, the population distribution of arctic charr is governed by this limitation of its eggs. Since its eggs will not hatch in temperatures above 20 degrees Celsius, arctic charr are found only in colder waters in the arctic. If you put adult charr in the warm waters of, say, the Gulf of California, they might live out their lives and spawn, but their eggs would not hatch, so the population would die out. Similarly, the saplings of various oaks and pine trees require different soils and light conditions. As a result, although the adult trees can tolerate wider variations, their distributions are limited by what their saplings can tolerate. Many living things are affected by Shelford's law, restricting their distribution and thus allowing space for a richer diversity of species on the planet.[102] Some animals deal with Shelford's law by constructing their own microhabitats with conditions that are salutary to their most vulnerable life stages. For instance, birds build nests and rabbits dig burrows for their young. Honeybees can persist in a variety of conditions, including cold climates, by creating a microhabitat inside their hives where their fragile eggs and larva can survive and grow.[103]

100. Krebs, *Ecology*, 35.
101. Ibid.
102. Dubos, *God Within*, 36.
103. For a delightful account of the lives of bees and many other insects, see Teale, *Strange Lives*.

Dusty Earthlings

At first glance, we humans appear to violate Shelford's law. As we have noted, humans live in or visit every ecosystem on earth. But in reality, we are delicate, fastidious creatures that require specific conditions in order to survive and thrive. Infancy is our most vulnerable life stage. As babies we require meticulous protection and care, as all experienced mothers can attest, but even as adults we can tolerate only narrow ranges of temperature, oxygen levels, humidity, salinity, acidity, concentrations of trace elements, and so on. Although we humans can become acclimated somewhat to higher or lower temperatures, we would not long survive wandering naked in the Sahara Desert or immersed in an icy lake with the arctic charr. To avoid such lethal conditions, like the birds, rabbits, and bees, we construct and encapsulate ourselves in microhabitats using technology such as clothing, houses, terrestrial and aquatic vehicles, heaters, air conditioners, and so on. Thus our immediate environment is maintained within the limits of our tolerance. Like many other species, we are especially attentive to our offspring. We carefully construct and maintain stable, protected environments for them that shield them from predators, competition, disease, variations in environmental conditions, and other members of our own species who might harm them. So Shelford's law still applies.

Our use of technology, energy, and resources to obey Shelford's law and maintain our microenvironments affects our ecological relationships.[104] For example, we use air conditioning to maintain cool microhabitats in hot climates. The widespread use of air conditioning has allowed the rapid growth of such cities as Phoenix, Arizona. In the long run, the large quantities of energy and resources required to do this may or may not be sustainable,[105] but the point here is that we cannot transcend Shelford's law. It circumscribes our lives and is an example of the limits that characterize our existence as eco-physical beings.

Similarly, there are limits on the harvesting of wild animal populations that we use for food. With rising global demand for fish, improved industrial fishing technology, and vast fleets of fishing ships and boats, worldwide fish populations have declined in recent decades. A 2003 analysis of worldwide fish catches suggested that global populations of large, open ocean predatory fish, such as tuna, marlin, and sharks, have

---

104. Odum and Barrett, *Fundamentals*, 179.
105. Holling and Meffe, "Command and Control," 187.

Ecology for Dusty Earthlings

declined by 90 percent since 1950.[106] Even if this study is only partly accurate, it indicates that we humans may be exploiting oceanic resources at a level beyond sustainable limits. This could lead to the collapse of some of these wild fish populations in the future. Marine biologists are improving their understanding of the ecology and dynamics of these fish populations, but they are difficult to study, and progress is slow. In the future, this knowledge may allow reasonable estimates of sustainable yields that could guide fishermen to limit catches so that stocks are not degraded.[107] But the point is that the harvesting of oceanic wild fish populations is not limitless as was formerly believed.[108]

Having said all this, we need to stop and take stock. The truth is that ecological limits exist, but they are complex, dynamic, and difficult to pin down. As we noted regarding population and carrying capacity, multiple factors interact to move limits up or down. It may be the case that today in the twenty-first century, we humans have reached sufficient numbers and sufficiently high levels of affluence and technological power that we are beginning to run up against some of the limits of the earth's capacity to support us. This is almost certainly true with respect to ocean wild fish populations, fresh water supplies, and climate change, but it is less clear with respect to other things such as land use and biodiversity. The important lesson is that *the principle of limits holds for all species, all ecosystems, and for the earth as a whole.* Our natural drive to obtain our basic needs for food, water, shelter, security, social support, and reproduction, as well as our drive to obtain wealth, power, and affluence, can impair our ability to recognize limits when we approach them and to modify our behavior accordingly.

106. Myers and Worm, "Rapid Worldwide Depletion," 282.

107. Odum and Barrett, *Fundamentals*, 301.

108. With decreasing yields from wild fish stocks and continued rising demand for seafood, fish farming or aquaculture has grown rapidly in recent decades. In 2004, about 27.2 million tons of marine products were produced by aquaculture, while in the same year, about 85.8 million tons of wild marine species were harvested. (United Nations Food and Agricultural Organization, "World Review of Fisheries and Aquaculture," http://www.fao.org/docrep/009/a0699e/A0699E04.htm#4.1.2). Fish farming may appear to be a solution to burgeoning world demand for seafood as wild fish stocks dwindle, but it is not without its problems and limits. Destruction of coastal wetlands and habitats to make room for fish farms, pollution, degradation of wild fish gene pools when farmed fish escape and interbreed, and the utilization of wild fish such as anchovies and krill as food for farmed fish all call into question the wisdom and sustainability of aquaculture as a solution to the overexploitation of the earth's oceanic food resources. See Naylor et al., "Effect of Aquaculture."

Dusty Earthlings

We humans are indeed remarkable creatures, and we do some remarkable things—both good and bad. But we are neither infinitely good, nor infinitely bad, nor infinitely intelligent, nor infinitely powerful, nor do we live in an infinite world. We are limited both from within and without. We are capable of amazing accomplishments and remarkable good, but we are also capable of remarkable failure and, on occasion, of stupendous evil. We can be thoughtful and wise, but we can also be mindless and foolish. And no matter how clever we think we are, the constraints and vagaries of earthly existence still circumscribe our lives—and our brains and our hearts. Jared Diamond's important book *Collapse* catalogues the histories of some human societies when they have succeeded or failed to obey the ecological realities of the various places where they have existed around the world.[109] The modern belief in human limitlessness and the limitlessness of earthly existence is false—scientifically, economically, and theologically. Finiteness and limits are inherent features of created existence for all creatures, including us. It is a basic "law of nature" that we humans must obey if we are to live in a Christian way on earth or hope for a livable future for our posterity and for our fellow creatures.[110]

The New Ecology

In the last section of this chapter, I want to discuss what has been called the "new ecology." In the second half of the twentieth century, there has been a shift in how ecologists understand ecosystems that is important for us laypeople to know about since it has implications for how we understand ourselves as part of those systems.

During the first half of the twentieth century, the dominant view of ecosystems was that they were stable, self-regulating systems that maintained themselves in "unity, balance, and harmony."[111] If left to themselves (that is, without human interference), ecosystems would remain stable and balanced over time. You have probably heard terms like "the balance of nature" that reflect this idea. "Undisturbed" wilderness was said to be "pristine" and "virgin," which meant that it had not been modified by humans and thus remained stable and healthy.

---

109. Diamond, *Collapse*.
110. Nash, "Christianity's Ecological Reformation," 374.
111. Callicott, "Multicultural Environmental Ethics," 79.

Ecology for Dusty Earthlings

This idea that the natural world involves balance, harmony, stability, and equilibrium in the absence of human "interference" has been widespread in popular ecological thinking. For instance, environmentalists have sometimes argued for the protection of "virgin" wilderness or for the preservation of the "balance of nature." The goal for us humans was to avoid upsetting things, to leave ecosystems in their "natural" state. In other words, we humans were not seen as part of nature. We were outsiders, so anything we did constituted external interference, which environmentalists condemned.[112] We were potential intruders and meddlers who needed to stay out and "let it be."

This view of nature as a balanced, harmonious, stable, finished product that it is our duty to let alone and let be has strong appeal for Christians. We can read Genesis as presenting creation as a finished work of God's perfect hand.[113] God designed and created it thus, and thus it should remain. We humans should not disturb it. Thus, Christian environmentalists have been perhaps more inclined to invoke this idea in support of ecological care than their secular counterparts. One encounters this idea not infrequently in Christian ecological writing even today.

But around the middle of the twentieth century, the thinking of ecologists began to change. Evidence was building that ecosystems were not stable, harmonious, and balanced but were ever-changing, dynamic, open systems. Flux, change, disharmony, variability, disturbance, and disequilibrium became the words that ecologists used to describe ecosystem dynamics. Fires, floods, disease outbreaks, droughts, and so on came to be seen as integral to the lives of many, if not all, ecosystems. As early as 1930, the great Canadian ecologist Charles Elton, studying animal populations, wrote, "The balance of nature does not exist, and, perhaps, never has existed."[114] "Most ecologists now reject any idea of a *balance* of nature, and the nonequilibrium paradigm is now the organizing principle of modern ecology."[115] Nonequilibrium means that ecosystems are dynamic and constantly changing. More recently, Christian geneticist R. J. Berry reiterated Elton's observation: "There is

---

112. Pickett et al., "New Paradigm," 68.

113. Keller and Golley, "Community," 107.

114. Charles Elton, *Animal Ecology and Evolution*, quoted in McIntosh, *Background*, 167.

115. Tarlock, "Nonequilibrium Paradigm," 1129.

almost certainly no such thing as a 'balance of nature.'"[116] So the idea that ecosystems are stable and unchanging has been abandoned.[117] In its place is the so-called *nonequlibrium paradigm* or *new ecology* of flux, change, variability, and disturbance.

But what is more important for us humans, and for the argument of this book, is that the new ecology has readmitted us humans back into nature. No longer are we regarded as outsiders who interfere in nature. Ecologists (and most environmentalists) today have come to see that *we humans are part of nature—integral parts of the ecosystem*. Archeological, historical, and ecological research has shown that humans have been intimately involved in nature for millennia—changing and shaping ecosystems and being changed and shaped by them. For instance, what were formerly thought to be pristine, undisturbed forests and grasslands of North America are now known to have undergone significant changes due to the activities of Native American peoples over millennia.[118] In fact, ecologists now talk about and study human-made cities and farms as urban and rural farm ecosystems (as I have been doing in this chapter).[119] The new ecology affirms my thesis that we humans are eco-physical beings, part of God's eco-physical creation.

There is a great deal that could be said about this topic, but space does not allow it. Nevertheless, I want to mention a few implications of the new ecology for us Christian laypeople.

(1) Change and Flux Are "Normal"

We humans never remain the same. We are always changing—growing, learning, improving, declining, maturing, aging, etc. Ecosystems are just the same. They are always changing. Understanding this is important. Since we rely on ecosystems to support us, we need to accept ecological change as "normal" and arrange our lives around it. We cannot keep ecosystems the same. We need to live within and manage them in such a way as to allow for change.

---

116. Berry, "Creation and the Environment," 32.

117. A word of caution here. Things are more complicated than we would like them to be. In reality, equilibrium and balance probably do play a role at many places and times in ecosystems. Thus, equilibrium and balance should probably not be discarded altogether. See Belovsky et al., "Ten Suggestions," 348.

118. Ibid., 192.

119. Odum and Barrett, *Fundamentals*, 359.

(2) Disturbances Are "Normal"

Disturbances—storms, fires, droughts, floods, outbreaks of disease, and so on—that stress or destroy ecosystems are normal. "These events are all part of the natural order. They have been recurrent, and many species are adapted to them. Natural disturbances can play a rejuvenating role in ecosystems, often increasing soil resources, plant growth, and reproduction of disturbance-dependent species."[120] In today's world, by far the most widespread and powerful agent of disturbance is humanity. We remove grasslands and forests and replace them with cities and farms; we introduce thousands of new chemicals into ecosystems every year; we transport and introduce nonnative species and diseases around the world; we extract millions of tons of fish from the oceans; and we release billions of tons of waste into the ecosphere each year.

At this point you may be saying, "But he just said that the new ecology takes humans to be part of the ecosystem, so why is he now citing all this 'damage' we are doing? If disturbances are normal, and we humans cause disturbances, then are we not acting normally?" Well, we *are* part of the ecosystem, and our lives and behavior *are* part of the dynamics of it. That is not the question. The question is, *How* are we to *be* part of the system and *what kinds and sizes of disturbances* should we cause? We are rational, ethical creatures with the capacity to learn about ecosystems and choose how we will live within them. If we cause disturbances, what, when, where, and how big should they be? We have made serious mistakes in the past because we didn't understand how ecosystems worked, nor did we think through carefully (or at all) what the effects of our disturbances might be. Furthermore, how we live and behave within the ecosystem must be guided by moral resources. For us Christians, these are Scripture, theology, and the Lord Jesus Christ. Only recently have we Christians begun to apply these to the Ecological Problem. We have a long way to go.

Our knowledge of disturbances and of ecosystems can help us design the disturbances we cause so that they are more likely to be helpful and less likely to do lasting damage. "Some human-caused changes may, in fact, mimic natural disturbances and support biological diversity."[121] For example, planning of cities and communities around rivers that

---

120. White, "Disturbance," 177.
121. Ibid., 192.

flood periodically could be done so as to take advantage of the fertility and renewal that the flood brings to the land. This is, in effect, what the ancient Egyptians did around the Nile. But if we cause too large a disturbance, it can damage an ecosystem beyond its ability to recover or remain productive. This has happened, for example, in the Sahel of Africa, where overgrazing and the cutting of trees have resulted in desertification—reduced rainfall and lower productivity—and it is happening in the United States, where good farmland is appropriated for suburban development. Ecologist Peter White has studied the mountain ecosystems of Smokey Mountain National Park in the southern Appalachians. He writes:

> Unfortunately, the scale and intensity of human disturbances often does indeed lie outside the bounds of natural variability and the evolved range of responses of the biota. Fire in the Smokies is a good example: fire in the 1920s developed on huge park clear cuts and, in this high rainfall region, resulted in so much post-fire soil erosion that some sites are still unforested seventy-five years after the fires. Clearly those fires were outside the natural range of variability in the park in terms of size and intensity . . . We must not accept the analogy between natural and human disturbance uncritically but must ask about the absolute nature of that change.[122]

We are part of the system, and we will cause disturbances. Humility, caution, and careful thought are required.

(3) Management Is Unavoidable

If ecosystems are complex and changing systems in which humans are integral parts, then our management of them is inescapable. We cannot "leave it be." Ecosystems inevitably change, so we need to try to find out how they change and try to influence them in directions consistent with scientific knowledge and our values and norms, and in ways that sustain human life and all of life. As Christian conservation biologist Fred Van Dyke says, active management is a "vital component to stewardship," and the causing of disturbances may be a "necessary and positive function of such management."[123] The new ecology throws us right back

---

122. Ibid.
123. Van Dyke, "Bridging the Gap," 162.

to the dominion mandate (Gen 1:26–28). As Scripture has instructed from the beginning, we *are* managers, overseers, caretakers, "stewards," or whatever you want to call it.[124] The question is not whether or not to manage the ecosystem; it is how to manage it, how to be a part of it, for the sake of what or who do we manage the system, and with what virtues and values.

(4) The End of Command-and-Control

We modern, wealthy, powerful Westerners are control freaks. We try to control everything we can. We want to make the world do what we want it to do, be what we want it to be, when we want it to be. This is part of what is called the "modern project"—the quest to master nature every day in every way. We want to eliminate inconvenience, uncertainty, variability, and unpredictability.

This approach has been used in ecosystem management. We dam, rechannel, and straighten rivers; we eliminate complex ecosystems and replace them with simplified monocrop agriculture; we try to eliminate "undesirable" species from ecosystems; we suppress fire in fire-prone regions; we build cities in resource-poor regions and then import massive quantities of resources to support them; we develop complex management mechanisms and bureaucracies in order to regulate nature, and so on.[125] The problem is that ecosystems don't work this way. They are "moving targets, with multiple potential futures that are uncertain and unpredictable."[126] The simple fact of the matter is that we will *never* be able to command-and-control God's creation or any part of it. The new ecology affirms the ancient biblical message that we are not God. God is God. He has appointed us managers but not masters. He is master, not us. He can command and control; we cannot. In fact, ecologists suggest that rather than trying to control ecosystems so that we eliminate messiness and irregularities, we should try to retain variation and diversity.[127]

---

124. I put scare quotes around the word "stewardship" because it is shrouded in controversy. See, for example, Berry, *Environmental Stewardship*.

125. Holling and Meffe, "Command and Control," 180–81.

126. Ibid., 183.

127. Ibid., 186.

Dusty Earthlings

This is a complex issue, and I don't mean to say that we shouldn't try to control nature at all. What I am proposing is humility, sensible moderation, and perspective. We should try to understand our place within the order of things; listen to ecologists when they tell us about how things work and what we should do;[128] seek to manage in a cautious way as best we can, realizing that things probably will not go as planned; be willing to learn and change; and trust that God is in ultimate control, not us.

Conclusion

The coast of Georgia is occasionally struck by hurricanes. Christian ecologist Susan Power Bratton recalls that in 1898, on St. Simons Island, Georgia, there was a severe storm that produced a twenty-two-foot storm surge on top of an eight-foot high tide. Thirty feet of water swept over the area, destroying everything less than thirty feet above sea level. Today there are summer homes built at the edge of the beach, only a few feet above the mean high tide line. A similar storm now would destroy all these homes and possibly result in considerable loss of life. Bratton comments, "We've lost our respect for wild nature... We think because we can place some rocks where they accumulate sand or can use heavy equipment to dredge out a channel, we can do anything ... or withstand anything."[129] Bratton asks, "Can we really hold nature by the tail and have her do our bidding without having her turn around and devour us?"[130] Our human arrogance and exceptionalism must be set aside. Despite our big brains, our technology, and our wealth, we remain dusty earthlings, subject to the ecological realities that God has written into his creation.

I have described a few of these realities that define and circumscribe our existence on this planet. These may not be as well understood as, say, the laws of physics, but we do understand them well enough to know that if we are not humble, cautious, and thoughtful, we may permanently destroy the very systems and creatures on which our lives depend. Hurricanes and storm surges occur along the Georgia coast. If we

---

128. MacIntyre, *Dependent Rational Animals*, 96–97.
129. Bratton, *Christianity, Wilderness, and Wildlife*, 108.
130. Ibid., 107.

build our house on the beach, a few feet above the high tide line, there is a good chance that sooner or later, our house (and maybe we too) will be annihilated. If we build our house in the hills behind San Diego, in the midst of the chaparral where periodic wildfires are part of the ecosystem, there is a good chance that sooner or later, our house (and perhaps we too) will be incinerated. If we build a huge city like San Diego in an area with few resources and then depend on fossil fuel–based supply lines to import almost all our resources, we are vulnerable to disturbances in those supply lines. Sooner or later such disturbances may occur, severely jeopardizing our fragile arrangements and our lives. If we continue to degrade land; if we continue to pump greenhouse gases into the atmosphere at accelerating rates; if we continue to overexploit water resources; if we continue to strive to expand our economy and to accumulate vast wealth in violation of the ecological limits of planet earth, then we, or our descendants, will, at some point, pay a price. We must consider too the poor people of the world, and the other species at whose expense we have built our modern edifice (Tower of Babel?) of power, luxury, convenience, wealth, and technology. We must learn to live *with* other people, *with* other animals, *with* the plants, *with* the microbes, *with* the land, *with* the seas, *in* the ecosphere of God's good earth that he has so graciously placed in our care. In the next chapter, I offer a Christian approach to this challenge.

7

Ethics for Dusty Earthlings: Living In God's Eco-Physical World

IF WE ARE DUSTY earthlings embedded in the earth's ecosphere, and we have a scriptural mandate to manage and take care of it under God's authority and to his glory, what should we do? How should we live? What kind of people should we be? These are questions about ethics—the fourth part of my argument (ch. 1, page 4, argument 4). In this chapter, I want to offer some ideas toward answering these questions. These are not "final answers." As I said in chapter 1, I do not claim to make any final statements about how things are and what we must do. But I do hope to stimulate some reflection, discussion, prayer, and maybe even action. The topic of ecological ethics is a huge topic that has filled many books. So I shall confine my discussion to four concepts that, I think, follow from what I have talked about. These include two virtues—*humility* and *self-control*—and two principles—*kenosis* and *justice*. Lastly, I shall discuss our joining with creation in the praise of God. I hope to show that in science, in Scripture, in our theology, and in Jesus Christ, we Christians have marvelous resources for developing and living as Christians in this beautiful world that God has graciously placed in our hands.

Ethics is the study of morality—right and wrong. It examines what kind of people we ought to be, how we ought to live, and the kinds of actions we ought to engage in. Most of the time, Christian theologians, philosophers, and ethicists talk about social morality, that is, morality that applies only to human affairs. But down through history a few have ventured outside the human realm into ecological ethics—how we should relate to the nonhuman part of creation: other animals, the land, the seas, and the earth. In the last fifty years or so, since the Ecological

Problem has become a public issue, there have been voluminous discussions of ecological ethics. For Christians the extension of ethics into the nonhuman world has been halting, wobbly, and contentious. But we are making progress. Herein is what I hope is a helpful contribution to the conversation.

"Enlightened Self-Interest" and the Fallacy of "Humans Versus Nature"

As I have mentioned, I was formerly a medical doctor, and for eight years (1985–1993) I worked as a medical missionary in the Comoro Islands (near the east coast of Africa) where malaria was a major public health problem. While I was there I treated many people with malaria and worked on a malaria control program. After my return to the United States, I had a discussion with a Christian friend about malaria in Africa. He had never been to Africa and had little knowledge of malaria but argued that since humans are far more valuable than anything else, nonhuman creatures and the ecosystem must be sacrificed for the sake of saving people from malaria. As I said, I treated hundreds of cases of malaria and studied the disease in depth. I told my friend that the epidemiology of malaria and the ecological interactions among humans, mosquitoes (the insects that transmit malaria), and plasmodia (the microbes that cause malaria) are complex. I said that prevention of malaria and control of mosquitoes needed to be integrated with other public health and development measures as well as being ecologically responsible—a holistic approach that incorporated the understanding that we humans are part of the ecosystem and dependent upon it. Mosquito control programs that damaged the ecosystem could have long-term consequences that might also hurt people who are part of the ecosystem. For example, the inappropriate use of insecticides would kill bees and other insects that pollinate crops, as well as predatory insects, spiders, birds, and reptiles that kill mosquitoes and other pests. This could hurt farmers and actually increase the spread of disease. But he would have none of this. He derided ecological concerns as superfluous. Citing the Golden Rule and our duty to care for the poor, he insisted that the only priority was people. My friend seemed to be assuming that humans and ecosystems were necessarily at odds with one another. To help people we must sacrifice nature; to help nature we must sacrifice people. For him, it was a zero-sum game: humans *against* nature.

A similar argument is made when economics and ecology are placed in opposition. During the 1990s, to protect dwindling populations of the northern spotted owl under the Endangered Species Act, the federal government took steps to reduce logging in the Pacific Northwest. A ferocious public debate ensued that was often framed as a conflict between ecology and economics—owls versus jobs. This seemed to be the underlying assumption of both sides—the loggers *and* the environmentalists. The truth of the matter was far more complicated than either side was willing to admit, but this controversy illustrates a frequent "fault line" in American culture that pits human economic interests against ecological concern.[1] Again, the issue was framed as humans *against* nature.

Seeing problems in this "humans versus nature" way assumes that humans are not part of creation or that we have different needs and operate in a different way, and that in order to insure human welfare, we have to sacrifice other creatures and ecosystems. It's one or the other. No other options exist. I suggest that this is a fallacy. Since we are actually part of the ecosystem, such "humans versus nature" thinking is wrongheaded. In principle, there is no opposition. What is in the interest of the ecosystem and its creatures is also in the interest of us humans and vice versa. All creatures are interrelated and interdependent. There is no us versus them.[2]

Now, the world is complex. Certainly, there are occasions when a choice must be made between human welfare and that of other organisms, or between one organism and another. In our malaria control program in the Comoro Islands, we took steps to reduce mosquito populations in order to protect people. In other words, we killed mosquitoes—or tried to. But our goal was not to obliterate the mosquito populations, and the methods we used to control mosquitoes respected the integrity and long-term sustainability of the ecosystem and the welfare of human communities and of other creatures. Or, at least, we tried

---

1. Speth, *Red Sky*, 107.
2. On this view, the common distinction in eco-ethics between anthropocentrism and biocentrism is muted. Human welfare and the welfare of the biotic community are aligned. For Christians, biocentrism and anthropocentrism should come together in a holistic theocentrism. See Van Dyke, *Between Heaven and Earth*, 195.

Ethics for Dusty Earthlings: Living In God's Eco-Physical World

to harm them as little as possible. We did try to kill mosquitoes, but we tried to do it in a responsible way. So sometimes it is necessary to kill or "sacrifice" other creatures for the sake of humans, but it should be done in a humble, thoughtful, and prudent way. A holistic, integrated view leads to a balanced approach.

Scott Sabin, director of Plant with Purpose, a Christian organization that helps poor people through agricultural development and ecological restoration, recognized early in his career that the welfare of people is bound up in the welfare of the ecosystem. He writes, "People frequently ask me how we can choose between caring for the poor and caring for creation, as if they are mutually exclusive. The question surprises me, because my own concern for the earth grew directly out of concern for the poor . . . Any response to the needs of poor people that hopes to be sustainable must consider the environment. Conversely, any sustainable conservation effort must consider the needs of the poor. *People and creation are part of the same system and are intimately connected. If you hurt one, you hurt the other.*"[3]

If we are eco-physical beings, of a piece with the earth and its natural systems, then we should approach all our problems with the assumption that, in principle, there is no conflict between humans and nature. Recognition of our eco-physical nature means that self-interest and ecological interests are (or ought to be) aligned.

This more holistic approach evokes what some have called a principle of "enlightened self-interest"[4] in which we care for ourselves by caring for the ecosystem and vice versa.[5] Eco-writer Wendell Berry comments:

> You cannot damage what you are dependent upon without damaging yourself . . . It is impossible, ultimately, to preserve ourselves apart from our willingness to preserve other creatures, or to respect and care for ourselves except as we respect and care for other creatures; and . . . it is impossible to care for each other more or differently than we care for the earth . . . The earth is what we all have in common; it is what we are made of and what

---

3. Sabin, *Tending to Eden*, 87–88; italics mine.

4. I put the phrase "enlightened self-interest" in scare quotes because I recognize its ambiguities and potential for misunderstanding.

5. Conradie, *Ecological Christian Anthropology*, 121.

we live from. Therefore, we cannot damage it without damaging those with whom we share it.[6]

We should not thoughtlessly sacrifice other creatures, ecosystems, land, air, water, or any of God's creation on the "crude belief that [humanity] is the only value to be considered."[7] As part of God's creation, in our human "self-interest," we should seek to align our lives and activities with the ecology of God's planet. Our utilization of resources and ecosystems for our needs should similarly be aligned with ecological realities.

But there are some caveats that, I think, need to be kept in mind if "enlightened self-interest" is to work as an ethical approach to our relations with the rest of creation. It should be conditioned by at least three principles. (1) We must have a deep understanding of ourselves as eco-physical beings—part of the ecosystem. Our characters, attitudes, lifestyles, and actions can only align with ecological realities insofar as we grasp our rootedness within God's ecological order. (2) We must accept and practice true Christian norms and virtues as ecological beings. This includes such things as humility, self-control, kenosis, and justice, which I shall discuss shortly, along with others, such as wisdom, prudence, sufficiency, frugality, love, and theocentrism.[8] In short, "enlightened self-interest" will only work if we seek to be more like Jesus in a concrete, material, ecological way. (3) We must have a more *holistic* and *longer* vision. We must overcome the tyranny of the urgent and our tendency to attend only to what is immediately before us.

I am not suggesting that the only way in which we value nature is for the sake of ourselves. As I noted in chapter 5, creation is theocentric. It is centered on God, not us. This means that ideally, we Christians will have a broader view in which we see creation as centered on God, and ourselves as part of God's "environment," and we think, live, and act always in reference to God—something we are not in the habit of doing, especially in things ecological.[9]

---

6. Berry, *Art of the Commonplace*, 111, 118.
7. Dubos, *God Within*, 45.
8. Bouma-Prediger, *For the Beauty of the Earth*, 137–60.
9. A corollary to theocentrism in Christian eco-ethics is *intrinsic value*, the notion that nonhuman creatures and things have value in themselves apart from humans. While this is clearly a biblical idea, its application in eco-ethics is challenging to say the least.

As biblical Christians, we can, I think, articulate and live a kind of "enlightened self-interest" if we understand ourselves deeply as dusty earthlings, if we seek to live more like Jesus, and if we think in terms of the broader picture and the long term—all for the sake of and to the glory of God. As Christian conservation biologist Fred Van Dyke has suggested, a reconciliation between humans and creation is needed in which the needs of humans and of other creatures are recognized and taken to heart in both attitude and action.[10] Understanding ourselves as part of creation helps us move in this direction.

Two Ecological Virtues

As eco-physical beings, what kind of people ought we to be? What *ecological virtues* might we seek to embody as individuals, as churches, and as communities? What virtues emerge from an understanding of ourselves as dusty earthlings, made in the image of God, who with creation are reconciled to God through the blood of Jesus Christ, and who seek to manage his creation in the way of Jesus? Ethicists such as Steven Bouma-Prediger have cited several of these, but I would like to focus on just two that, I believe, are important in helping move us toward living more ecologically and Christianly within God's creation. These are *humility* and *self-control*.

Humility[11]

Humility has been defined as having an accurate understanding of who and what we are and of our place within the order of things. In the Bible, humility is carried further and speaks of lowliness, poverty, and self-effacement.[12] Today it is a much needed virtue in many areas of life but especially in our ecological relationships.[13] In a sense this book amounts to a plea for humility in us Christian humans. A genuine humility is, in my view, essential to any viable eco-ethics. We all make the mistake, at one time or another, usually unconsciously, of placing

---

10. Van Dyke, *Between Heaven and Earth*, 195.
11. Bouma-Prediger, *For the Beauty of the Earth*, 146.
12. Bauer et al., BDAG, 981–90.
13. Bauckham, *Bible and Ecology*, 46.

ourselves rather than God at the center of things. This mistake has engendered pride and self-absorption and contributes to the Ecological Problem. Humility is a basic biblical virtue. It is expressed in the creation narratives, the flood story, the Lord's speech to Job, the preaching of the Hebrew prophets, the incarnation of Christ, and the paradigm of repentance. God calls us to assume our proper place within the created order, among the creatures—bone and blood amidst the dirt and rocks of God's good earth.

A proper humility is aided by an authentic embrace of ourselves as bodies, of our physical nature, of our fragility and vulnerability, of our place in the physical creation and our dependence upon it.[14] Humility should come from both a scientific and a theological understanding of ourselves. Scientifically, we understand ourselves as creatures—material beings—living in solidarity with the rest of creation. Theologically, we understand ourselves as creatures gifted with power over creation that we are capable of utilizing in the image of God—nurturing and making room for other humans and for other creatures just as God makes room for us.[15] We see creation itself, and ourselves within it, as a contingent gift to us, which we do not deserve. All of this provides a powerful warrant for a grateful and liberating humility.

In the face of our chronic arrogance, God himself commits the supreme act of humility. In Christ, God enters his creation and becomes a creature, an animal among animals—a human—suffering, bleeding, dying, just like other creatures (Phil 2:6–11). In this consummate act of humility (or *kenosis*, as I shall discuss shortly), we learn that the virtue of humility is bound up in the very nature of God himself.[16] God is love (1 John 4:8, 16), and he is also humility (John 1:14; Col 2). This *humility of God* is at the core of Christian eco-ethics, and of all of Christian ethics, for that matter.[17] In this action God shows us that humility means more than merely an "accurate and realistic understanding of who and what we are and of our place within the order of things." Divine humility is God's intentional, voluntary self-emptying, mortification, self-abasement, and sacrifice. In the incarnation, passion, and death of

---

14. Dyrness, *Earth Is God's*, 116.
15. Moltmann, *God in Creation*, 88, 156.
16. As I discussed in chapter 4, God's self-giving love is also manifested in his great act of creation.
17. Polkinghorne, "Kenotic Creation," 106.

Christ, the highest assumes the position of the most lowly. Hence, for us Christians, who understand ourselves as made in the image of this God of humility and capable of living out that image, humility extends to the willful practice of self-giving and self-sacrifice for the sake of others and for the sake of God's creation. This is the way of Jesus—the way of God.

But besides our theology and the person of Jesus Christ, concrete ecological experience should also teach us humility. In April 2010, the Deepwater Horizon oil platform exploded and sank off the coast of Louisiana in the Gulf of Mexico. The subsequent oil spill released an estimated 200 million gallons of oil into the gulf, causing extensive damage to the ecosystem and to economic interests in the region. This event, and others like it, should be a wake-up call for us humans—that despite our self-confident dominance over nature, our own fallibility and the uncertainties of earthly life should make us humble. Despite our prodigious efforts toward command-and-control over nature, we still make mistakes, and those mistakes can be costly. The fisheries and other industries of the Gulf Coast depend on the health of the coastal ecosystems. Humans are *embedded* in those ecosystems. The some four thousand oil rigs operating offshore in the gulf created a situation where human error and the inherent stochastic nature of creaturely existence would lead to a damaging spill sooner or later—and it did. No matter how smart, powerful, and perfect we think we are, our ignorance and fallibility will be expressed, and nature itself will outsmart and outpower us. A proper humility coupled with godly prudence, reverence, caution, love, and gratitude are called for. These are spiritual and moral virtues of human life that do not come from technology, education, government, or business. They come from God. The gulf oil spill is the responsibility of all of us. It is our own uncritical participation in the ways of the world (1 John 2:16–17), our overweening demand for oil, that creates the context for such expensive, risky, and technologically challenging human enterprises. In the end, as Susan Bratton says, "Creation itself teaches human limitations."[18]

Finally, divine humility ought to lead us to make room for other people and creatures in this world. As I discussed in chapter 6, we, the wealthy of the world, are taking far more than our share of the earth's resources and producing far more than our share of its pollution. God's humility in creation and in the redemptive work of Christ made room

---

18. Bratton, *Christianity, Wilderness, and Wildlife*, 92.

for us. Gratitude for this and the desire to imitate the God whom we worship ought to motivate us to intentionally make room for others in the world. This would suggest voluntarily reducing our consumption and our lifestyles. Besides, to be wealthy in the way of Jesus is to be poor in the way of the world (2 Cor 8:9).

Voluntary Self-Control (Self-Limitation)[19]

A virtue that is related to humility is *self-control*, or what the Apostles Paul and Peter called *self-limitation* or *self-restraint*.[20] Some authors have written of the closely related virtues of frugality, temperance, and moderation.[21] What I am talking about here is actually limiting ourselves—not doing things we have the power (and money) to do—materially, concretely, voluntarily, freely, cheerfully—not because of circumstances, laws, or economics, but because we trust God, we have grasped the grace of God, and we genuinely understand that self-limitation is integral to authentic faith and to the health and flourishing of ourselves, of other humans, and of all God's creation. As Francis Schaeffer wrote back in 1970, "So I am to put a *self*-limitation on what is possible. The horror and ugliness of modern man in his technology and in his individual life is that he does everything he can do, without limitation. Everything he *can* do he *does*. He kills the world, he kills mankind, and he kills himself. I am a being made in the image of God. Having a rational-moral limitation, not everything man can do is right to do."[22] Self-limitation is saying, "No, I have enough" when the world says, "Take more." In a real sense, self-control is the living out of humility. The conscious, material practice of intentional self-limitation embodies concretely the image of God in us and contributes to the welfare and flourishing of ourselves, others, our posterity, and all of God's creation. Although this virtue is embodied supremely (and physically) in Jesus Christ and is well known to the writers of the New Testament, it is little known to our modern, hyper-developed consumer culture whose values of self-indulgence, entitlement, consumption, and the relentless

---

19. In New Testament Greek, *enkrateia* means "restraint of one's emotions, impulses, or desires, self-control" (Bauer et al., BDAG 2000, 274).

20. Acts 24:25; 1 Cor 9:24–27; Gal 5:23; Titus 1:8; 2 Pet 1:6.

21. Bouma-Prediger, *For the Beauty of the Earth*, 144–46.

22. Schaeffer, *Pollution*, 90–91.

pursuit of wealth are contrary to the values and virtues of our Lord and of his kingdom.

Voluntary self-limitation requires *personal autonomy* in that we consciously resist the consumer culture and economic forces (the advertising and marketing industries) that seek to manipulate us to their own advantage. We are continually bombarded with explicit and implicit messages designed to induce buying, eating, and consuming behavior.[23] An example is Christmas, which has become commercialized far beyond all reason (as have Halloween, Thanksgiving, Valentine's Day, and Easter). So self-limitation in the sense that I am suggesting here would involve the practice of personal autonomy and independence from these powerful, worldly forces of our culture. Just as the onslaught of the advertisers is relentless and unending, so also our practice of personal autonomy and self-control must be relentless and unending—with God's help.

In chapter 5, I argued that since we are part of creation, our dominion over creation includes *dominion over ourselves*. As theologian Hendrikus Berkhoff rightly observed, this is by far the most difficult part of the dominion mandate.[24] Ruling and subduing ourselves with gratitude to God, in the power of the Spirit, to God's glory, for the sake of others ought to engender within us Christian believers, as compared to nonbelievers, an observable (measurable) inclination toward material moderation and restraint. After all, we have the grace and example of Jesus to guide us. Unfortunately, most of us never really articulate these realities, let alone put them into practice. As a result, we Christians live more or less like everyone else does. We have the same amount of stuff (and of debt), and our desire for more is as strong or stronger. This lack of self-control is probably the manner in which in our everyday lives we most clearly abrogate the dominion mandate. God's gift of power to control the earth and its creatures should be joined with the power to control ourselves (Gal 5:23; 2 Pet 1:6).

Susan Bratton remarks that "by not ruling our own appetites, we ultimately cease also to rule creation."[25] Our uncritical participation in the excesses of our consumer culture is in reality an abdication of

---

23. Stassen and Gushee, *Kingdom Ethics*, 426.

24. H. Berkhof, *De mens onderweg. Een Christelijke mensbeschouwing*, quoted in Conradie, *Ecological Christian Anthropology*, 203.

25. Bratton, *Six Billion*, 11.

our dominion role, a negation of our freedom, and ultimately a form of enslavement. Self-limitation is a countercultural idea. It embodies "perhaps the most unnerving norm in North American societies," suggesting "thrift, moderation, temperance, and material sufficiency,"[26] virtues that are not encouraged in our culture. I have avoided using the word *law* for the ecological principles that I have discussed in this book. But I will here venture to offer a "spiritual law" of human ecology: *The one who cannot rule himself will himself be ruled*—and the unwelcome external ruler will be either nature itself, economic reality, or the government.

We Christians ought to be critics of consumer culture and the pursuit of wealth. The church ought to play a prophetic role here. Jesus turned human values on their head; he did not live in a palace but "had no place to lay his head" (Matt 8:20); he preached and lived selfless love and chose the way of the cross rather than the way of power and wealth (John 18:36). We modern, wealthy Christians live with the embarrassing fact that Jesus Christ, the founder of our faith, lived in poverty and in his preaching mounted a forceful critique of wealth.[27] The pursuit, accumulation, and maintenance of wealth is the engine of the form of capitalism we practice in the world today, a pursuit that is an exalted value of our culture and to which many, if not all, of us Christians are committed. In reference to Jesus' comments about birds, lilies, and our anxieties (Matt 6:25–34), New Testament scholar Richard Bauckham writes:

> Recalling our contemporary situation, we may now be able to see that Jesus' teaching is not a matter of economic naiveté or irresponsibility as has sometimes been alleged. Rather, it invites us to see the world and ourselves as God's creation and the resources on which we depend to live as God's provision for all his creatures. Certainly, Jesus' concern with subsistence contrasts sharply and instructively with the modern talk of wealth creation. At least in Western industrial society, the

---

26. Nash, "Ecological Reformation," 11.

27. Matt 5:5; 6:19–34; 13:22; Mark 10:17–31; 12:41–44; Luke 1:53; 6:20–25; 7:18–35; 12:13–21; 16:13–15, 19–31. In addition, other teachings and actions of Jesus support his critique of wealth: Matt 13:44–46; 20:1–16; Luke 6:30, 34; 14:33; 19:1–10. Several passages elsewhere in the New Testament support Jesus' critique of wealth: 1 Cor 1:27–31; 2 Cor 8:2; 1 Tim 3:3; 6:6–12, 17–19; 2 Tim 3:2–3; Jas 1:10; 2:5–7; 4:13–17; 5:1–6.

> instinctive human anxiety about having enough to survive to which Jesus refers has long been superseded by the drive to ever-increasing affluence and obsessive anxiety to maintain economic growth . . . Western people today are obsessively worried not even about maintaining the standard of living we have, but about maintaining its constant improvement. It is this anxiety that is depleting and destroying the resources of nature and depriving not only other species but many humans of the means even of mere subsistence.[28]

Bauckham here correctly identifies our modern pursuit of ever increasing wealth as an underlying cause of the Ecological Problem. We affluent Christians alternately berate and justify ourselves for our wealth, but we rarely, if ever, change our ways. Material, voluntary self-limitation is clearly embedded in the gospel of Jesus Christ. It is linked to our trust in God, to worship, and to authentic discipleship. For us, the wealthy Christians of the world, it will be a required virtue if we are to address the Ecological Problem in a meaningful way.

Wealth, of course, is not inherently evil; it is good. We all need *some* wealth in order to live. The question is not wealth versus poverty. It is how much wealth is needed? It harks back to the problem of moderation in carrying out the dominion mandate that I mentioned in chapter 5. When is enough enough? While more guidance from Bible scholars, theologians, ethicists, and pastors regarding how Jesus understood wealth and how his values could be lived out in our modern context would be helpful, wealth remains a very difficult problem for all of us. "No aspect of Jesus' teaching is so confrontational and so difficult to implement as his teaching on money."[29] The gap between Jesus' material way of life and my own, between Jesus' values and my own, is for me a source of pain. I will continue to struggle with wealth and with self-limitation and restraint. I hope you will too. If we do, maybe for both of us, someday, true repentance will come, and we will give up our possessions (not just "in our hearts" but in reality), choose the way of life in Jesus (John 14:6), find "true riches" (Luke 16:11), and discover not only eco-physical healing but also personal joy, peace, and liberation.

---

28. Bauckham, "Jesus, God and Nature," 222.
29. Snodgrass, "Jesus and Money," 135.

Eastern Orthodox Christianity contains a tradition of *asceticism*.[30] The Western church has also had an ascetical tradition but has been more ambivalent about it, especially in modern times. Today for us modern American Christians, asceticism has a negative connotation.[31] Practicing willful self-denial as part of the Christian life seems legalistic, foolish, and unhealthy. After all, for us Americans, "the pursuit of happiness" is an inalienable right. But we should remind ourselves that Jesus himself seems to have suggested asceticism, or something like it, when he said, "If anyone would come after me, he must deny himself and take up his cross and follow me" (Mark 8:34). Our Eastern Orthodox brothers and sisters may have something to teach us here. A little practice of self-control in the Christian ascetical tradition might help us toward deeper devotion to Christ, more authentic discipleship, and more ecological living. The current Ecumenical Patriarch of the Orthodox Church Bartholomew II says that "the traditional Orthodox virtue of asceticism" or "self-discipline" is "essential if humanity [is] to correct its flawed attitude to nature."[32] He's right. But again, asceticism, or self-limitation, is not something that we modern American Christians are attracted to or want to talk about.[33]

Christian theologian Loren Wilkinson offers *resistance* as a way of living out the virtue of self-limitation in response to modern consumer culture. In the face of the relentless pressure of advertising and consumerism, he suggests that we stand our ground and resist by turning off the television; reducing our visits to malls and to fast food restaurants; keeping that laptop around for a few extra years; driving a smaller car (or walking, riding a bike, or using public transit); limiting our long-distance travel; living in a smaller house or apartment—even if one has the means for something larger; and resisting the urge to have the

---

30. Asceticism is the practice of self-denial and renunciation as a spiritual discipline See Staats, "Asceticism," 131–34.

31. Dietrich Bonhoeffer writes that modern Protestants are unable to understand the practice of asceticism and other spiritual disciplines (*Ethics*, 297).

32 Cited in McGrath, *Re-Enchantment*, 39.

33. There is an embryonic movement among American Christians that generally goes by the rubric "simple living," a term more acceptable to Americans than "asceticism." These folks seek to gently push back against affluence, consumerism, and materialism, seeking a simpler, more holistic life that is focused on relationships, contentment, and sharing. For more information, simply type "Christian simple living" into your search engine and you will find a wealth of sources.

"latest and greatest" products and fashions. Wilkinson says, "In few other areas can we better demonstrate 'good living' and our allegiance to 'the living God' than by refusing to be shaped by our consumerist culture."[34]

Wilkinson is correct, but in itself, resistance may not be sufficient. What is needed is something positive, something framed in terms of the gospel, to help us transform our values and attitudes about what makes the "good life" and what is "true wealth" for us dusty followers of Jesus. If we rethink human "goods" in terms of the gospel, we might join with the Apostle Paul and consider all things "rubbish" compared to the surpassing glory of knowing Christ (Phil 3:8). Perhaps Paul offers Christ himself in place of wealth or "the pursuit of happiness." If we focused more on Christ, and our values were more like those of Jesus, then maybe "an eco-friendly consumption would not involve a reduction of living standards, but rather an altered conception of the standard itself."[35] In this regard, theologian Glen Scorgie points to the wonder and riches that are ours in the study and comtemplation of God and the great truths of our theology.[36] Besides that, God has given us his wondrous creation to appreciate and wonder at. Perhaps for us Christians the "good life" or the "American dream" or "happiness" or "wealth" would be defined not by our personal possessions or the values of the world, but by our connection with Jesus Christ and the values of the kingdom of God, and as a result we might live in a noticeably different way from that of the world around us. What the world views as poverty we might view as wealth, and what the world viewed as wealth we might view as poverty (Luke 16:15). We might find ourselves favoring and even involved in "new social and economic relationships that were more just and humane."[37] Jesus Christ, in his life and preaching, seems to point to a "good life" that, if it were lived, would align well with ecological realities and a reduced ecological footprint. If we were ever able to get it into our heads and hearts what the "good life" actually is for us followers of Jesus, we might move closer to living more ecologically, and our witness and evangelism might become much more powerful.

---

34. Wilkinson, "Eco-Myths," 31.

35. Kate Soper, *What Is Nature?*, quoted in Moo, "Nature in the New Creation," 488.

36. Scorgie, "Wonder," 24.

37. Metzger, *Consumption*, 196.

Dusty Earthlings

As with all virtues, there is a pitfall here. Self-limitation can become an end in itself and a source of pride. As Christian environmentalist Dave Bookless warns, we can become "eco-Pharisees," with a "greener-than-thou attitude in place of a holier-than-thou attitude."[38] He suggests that the best way to avoid this is to maintain a "simple, humble, childlike relationship of trust in, and dependence on, Jesus,"[39] which, after all, is what we are supposed to do anyway (Mark 10:15). Besides, Jesus was clear about not judging others.[40] Being obedient to Christ, humbly recognizing our ignorance of the lives, motives, and situations of others, and fixing our eyes on "Jesus the author and perfecter of our faith" (Heb 12:2) should help us avoid this pitfall. As social beings, we need the support, teaching, and correction of the church, if the church is truly oriented toward Jesus and the kingdom of God as it should be. Honest engagement with Jesus Christ, and openness to the counsel and correction of others (who are also engaged with Jesus), should help keep us on track.

Two Ecological Moral Principles

Kenosis

The principle of kenosis is a core idea in Christian theology, ethics, and life. As we saw in chapter 4, it means *self-emptying*, the giving up of position, power, and privilege, and it is integral to God's character. It is closely related to the virtue of self-limitation that I just discussed. In the incarnation, God the Son, who is infinitely abundant, emptied himself to become a finite human creature. Christ was fully God and fully human, but in his act of kenosis, he declined "to draw on his divine abilities merely at his own whim."[41] As a human, Jesus possessed divine capacities but chose not to use them or to use them only in limited ways, such as the raising of Lazarus and the calming of the storm.[42] Being both divine and human, Jesus made a continuous conscious choice

---

38. Bookless, *Planetwise*, 120.
39. Ibid.
40. Matt 7:1–5; Luke 6:37–38.
41. Grenz, *Theology*, 307.
42. Mark 4:35–41; John 11:38–44.

not to do things and not to use powers that he could have used. He limited himself in order to redeem the world that he loved. It is on account of this basic feature of God's character that Scripture says, "God is love," insofar as any single word or idea might describe God.[43]

In chapter 4, I emphasized the radical difference between God and us humans. We humans are not God. But in Jesus Christ we are presented with God's challenge to us to be like God in Jesus Christ as dusty earthlings. We are called to be the kenotic species. We are the only living thing on earth capable of choosing our way of life and our actions.[44] We are the only creatures with capacities for self-transcendence and for kenotic love. Kenosis is "a joyous, kind, and loving attitude that is willing to give up selfish desires and to make sacrifices on behalf of others for the common good and the glory of God, doing this in a generous and creative way, avoiding the pitfall of pride, and guided and inspired by the love of God and the gift of grace."[45] We could say perhaps that among all the creatures of earth, "the uniqueness of humanity consists in [our] ability to become a servant species."[46] Jesus Christ is the divine archetype of this servant species, and we Christians are (or should be) following in his way (Rom 8:29). As New Testament scholar Douglas Moo writes, "From a biblical-theological perspective, human dominion over creation must also be interpreted christologically. Christ's own sacrificial 'rule' provides the ultimate model for our own rule of the earth."[47] In the image of Christ, we are to be servant-rulers not only of other humans but of all God's creatures.[48]

Justice

Justice is a big topic, but part of what it means is that resources should be distributed equitably among people—all people. Everyone on earth ought to have access to resources such that their basic needs are met and they can live a decent life. Furthermore, our eco-physical nature and our embeddedness in the ecosphere mean that we are willy-nilly

43. Peacocke, "Cost of New Life," 41.
44. Rolston, "Kenosis and Nature," 63.
45. Ellis, "Kenosis as a Unifying Theme," 108.
46. Linzey, *Animal Theology*, 57.
47. Moo, "Nature in the New Creation," 479.
48. Van Dyke, *Between Heaven and Earth*, 196.

connected ecologically to other people on the planet, even people who live far away. What we do affects them, and what they do affects us. Ecologically, we cannot live in isolation from the broader world; all is interconnected.

As I noted in chapter 6, we Americans and other affluent peoples are taking more than our share of the world's bounty. "Over the past 40 years, consumption of ecological resources, per capita, has increased by nearly 90% in high-income countries but fallen by 15% in low-income countries."[49] There are numerous examples of how this takes place. Gold is mined in central and western Africa by laborers under miserable and dangerous conditions; some of this gold ends up in the jewelry stores of our shopping malls. Tropical rainforests are being cut down for a number of reasons, but the demands of affluent cultures in North America, Europe, and Asia for wood, meat, and other products helps drive the destruction. The rainforests of Borneo are being removed to make way for palm tree plantations that produce palm oil, which is used in soaps, cosmetics, cake mix, cookies, candies, ice cream, glues, inks, candles, and a host of other consumer products for rich societies.[50] The United States imports about one million barrels of oil from Nigeria every day. The extraction of oil in Nigeria is destroying the homeland of the Ogoni people, who are receiving none of the profits and are ruthlessly suppressed by their government when they protest. The greed of government officials, attracted to lucrative profits generated by America's voracious appetite for oil, leads to the oppression and destruction of the Ogoni people.[51] Our demand for paper, furniture, wood products, coffee, rubber, sugar, soybeans, meat, and a multitude of consumer goods removes ecological resources from poorer nations in order to supply the wants and whims of the wealthy. Affluent societies are producing the bulk of the greenhouse gases that are driving global climate change. If the predictions of climate scientists prove true, those who will suffer most will be the poor and vulnerable of the world, not the rich who are primarily responsible for problem.[52]

---

49. Wackernagel and Kitzes, "Ecological Footprint," 1035.

50. The destruction of the rainforests of Borneo is also reducing the habitat range of the Borneo orangutan, an endangered species.

51. Valerio, *Lifestyle*, 49.

52. Patz and Kovats, "Hotspots," 1094.

These are just a few examples of the ways in which wealthy societies are participating in ecological injustice and the oppression of the poor in the world. The gold mines in Africa and the destruction of the rainforests in Borneo degrade the land, reduce biodiversity, pollute rivers, and reduce the potential productivity of the land for local peoples. It is wrong to damage our neighbor's house in order to enhance our own, especially when we already have so much more than our neighbor. We are reminded of Nathan's rebuke of David, after his adultery with Bathsheba and murder of her husband Uriah. When Nathan confronts David, he evokes the idea of justice (2 Sam 12:1–12). David was king, yet he did not voluntarily limit himself; he thoughtlessly indulged in the pleasures of the consumer culture of that day and time. This stands in stark contrast to Uriah's honorable self-denial for the sake of others (2 Sam 11:11). Furthermore, the biblical ideas of divine creation and the common origin of humanity mean that everyone has a right to a reasonable share of the earth's resources.[53] In our interconnected global ecosphere, our neighbors are not simply the people who live next door or the homeless person on the street. Everyone is our neighbor. How we live—our lifestyle—matters; it impacts people around the world.

If we read our Bibles carefully, we find that biblical justice is "preoccupied with the needs of those who are poor, weak, disadvantaged, or oppressed."[54] Christian character should include a godly concern for the poor and a strong sense of justice that bears fruit in action and a changed life through the virtues of kenosis and voluntary self-limitation. As Pope John Paul II put it, "The earth is ultimately a common heritage, the fruits of which are for the benefit of all . . . It is manifestly unjust that a privileged few should continue to accumulate excess goods, squandering available resources, while masses of people are living in conditions of misery at the very lowest level of subsistence. Today the dramatic threat of ecological breakdown is teaching us the extent to which greed and selfishness—both individual and collective—are contrary to the order of creation."[55] We Christians ought to be concerned about this. Mahatma Gandhi expressed a Christian attitude when he said, "With regard to possessions, a person's goal should be to see how much he/she can do without. Whatever persons have

53. Valerio, *Lifestyle*, 34.
54. Stassen and Gushee, *Kingdom Ethics*, 70.
55. John Paul II, "Peace with God the Creator."

Dusty Earthlings

beyond their needs is stolen from the poor."[56] Christian ethics requires "the willingness to share economic goods."[57] In our globalized world, sharing means sharing globally. Justice is justice for all.

Praising God with Creation

> And God saw all that he had made, and it was very good (Gen 1:31).

> Let the sea resound, and all that is in it;
> let the fields be jubilant, and everything in them!
> Then the trees of the forest will sing,
> they will sing for joy before the Lord,
> for he comes to judge the earth (1 Chr 16:32–33).[58]

The Bible describes a reciprocal relationship between God and his creation. As we observed in chapter 4, God delights in his creation, and, as shown in the verses above, it offers praise back to him. God looks upon his creation and finds it "good, lovely, pleasing, and beautiful."[59] It is valuable and pleasing in his eyes, and he finds joy and satisfaction in its existence—in the diversity and beauty of living things, from the wild ass to the soaring eagle.[60] And creation beholds, praises, and glorifies its creator. The Westminster Shorter Catechism says, "The chief end of [humans] is to glorify God and enjoy him forever."[61] Bible scholar Chris Wright suggests that this is also true of the whole creation—it "exists for the praise and glory of its creator God."[62]

Since we humans are part of the creation, one of God's creatures, he delights in us too. And Psalm 148 calls us humans to praise along

---

56. Quoted by Jegen, "Church's Role," 95.
57. Murphy and Ellis, *Moral Nature*, 251.
58. Other passages that speak of creation's praise of God include the following: 1 Chr 16:30–33; Pss 19:1–6; 69:34; 93:3–4; 96:11–13; 98:7–9; 103:22; 145:3–7; 148; 150:6; Isa 6:3; 42:10; 44:23; Phil 2:10; Rev 5:13.
59. Brueggemann, *Genesis*, 37.
60. Bauckham, *Bible and Ecology*, 51.
61. Center for Reformed Theology and Apologetics, "The Westminster Shorter Catechism."
62. Wright, "'Earth Is the Lord's,'" 223.

Ethics for Dusty Earthlings: Living In God's Eco-Physical World

with all the other creatures.[63] Theologian Stanley Grenz writes that "glorifying the Father in the Son together with all creation is the ultimate expression of the *imago Dei* and therefore marks the *telos* [purpose] for which humans were created in the beginning."[64] In other words, it is our purpose as humans to glorify God the Father along with all of creation. Richard Bauckham writes, "The most profound and life-changing way in which we can recover our place in the world as creatures alongside our fellow-creatures is through the biblical theme of the worship all creation offers to God."[65] It may offend our pride to be listed with "hills, horses, and hurricanes" in Psalm 148,[66] but there it is in Scripture. We are (or should be) part of creation's praise of God, and it should be a source of continuous joy, delight, and blessing for us.

How can we do this? How can we join in creation's praise of God? I suggest there are at least four ways. First, our worship of God in song, music, praise, poetry, and prayer is part of creation's praise of God. The manifold forms of worship among Christian churches around the world harmonizes with the rest of creation in glorifying and honoring God. In this way, we are already praising God with creation. This praise is a vital and biblical way of bringing glory to God.[67] No change is needed except perhaps some encouragement that we participate regularly in the corporate worship of our churches.

The second way is less familiar to us. In 1970, theologian H. Paul Santmire wrote that a key role for Christians as "responsible citizens of the Kingdom of God . . . is in many ways the most attractive yet the least understood authentic human relation to nature. This is the divine call to be nature's *wondering onlooker*."[68] God intends for us to be *wondering*

---

63. Bauckham, "Joining Creation's Praise," 45.

64. Grenz, *Social God*, 327.

65. Bauckham, *Bible and Ecology*, 76.

66. Fretheim, "Nature's Praise," 16.

67. Christian music and songs also call inanimate creation and nonhuman creatures to praise God: the doxology "Praise Him All Creatures Here Below," the hymns "All Creatures of Our God and King," "This Is My Father's World," "Fairest Lord Jesus," and "Holy, Holy, Holy," and the praise song "Glorify Your Name in All the Earth." Other literature and poetry that calls creation to praise God include the "Benedicite," or "Song of Creation," taken from the Apocryphal book *The Prayer of Azariah and the Song of the Three Jews* (vv. 35–65), which has been used in Roman Catholic, Lutheran, and Anglican worship for centuries.

68. Santmire, *Brother Earth*, 190.

onlookers of all that he has made, *joyful appreciators, astonished students of God's handiwork* (Gen 2:19; Ps 8:3). Since we are made in the image of God, we can image God by *imitating his enjoyment of his creation.* Pastor Scott Hoezee writes, "A major part of our Christian vocation should be the nurturing of delight in this universe of wonders—a delight similar to God's own playful joy in creation."[69] God offers us a "natural world of abundance and beauty, which exists by the creator's gift, independently of all our efforts to create our own world of plenty and beauty for ourselves."[70] Not only does he save us from our sins, but he has created a marvelous biosphere, with amazing creatures, beautiful landscapes, and wondrous processes and phenomena. God has lavishly poured out his beauty, wisdom, and grace in everything from tiny flowers to mighty mountains, from an elegant shrub growing in the yard to a majestic sequoia on the slopes of the Sierra Nevada, from the intricate structure of a bacterium under the microscope to the fearsome power of a storm, from the gentle kiss of an autumn breeze to the delicate symmetry of a snowflake. The massive profusion of all of this speaks of the extravagant creativity and prodigal love of God. The great John Calvin wrote, "Wherever you turn your eyes, there is no portion of the world, however minute, that does not exhibit at least some sparks of beauty; while it is impossible to contemplate the vast and beautiful fabric as it extends around without being overwhelmed by the immense weight of glory."[71]

God has uniquely equipped us humans to experience these things—to see, hear, touch, smell, admire, wonder, enjoy, and learn. And God seems to have a sense of humor—a kind of playfulness. Insects, for example, take many forms that can make us laugh—and perhaps God laughs too. As Edwin Teale says in his delightful book *The Strange Lives of Familiar Insects*, "Imaginary wonders . . . are less needed in dealing with insects than with any other group of living creatures. The truth is odd and dramatic enough."[72] We Christians should be amateur biologists, geologists, botanists, zoologists, ecologists, astronomers—the ones, who like children, are forever pointing to the wonders around us, saying, "Look there! Look here! Wow! Praise be to God!"

69. Hoezee, *Remember Creation*, 8.
70. Bauckham, "Jesus, God and Nature," 222.
71. Calvin, *Institutes*, 1.5.1, p. 51.
72. Teale, *Strange Lives*, 3.

We can praise God by studying and learning about his creation and his creatures. The natural world has much to teach us if we are willing to take the time.[73] "We are called to wonder in the basic, religious sense, We stand in awe of a 'power' greater than ourselves as we contemplate the infinite space and the teeming complexity of life that surrounds us and has surrounded humankind from 'the beginning.' So overwhelming is the invitation to wonder that one wonders (in another sense of the word) if we are really capable of penetrating to a worthy theology of nature. We 'wonder' whether we have the resources, the patience, and the receptive hearts and minds required to penetrate and to understand adequately the world of nature about us."[74] Christian biologist Fred Van Dyke notes that "nature is a source that illuminates the glory of God," and "science illuminates nature."[75] In other words, the more we learn about and understand God's nature, the more we will understand the glory of God. Our study and contemplation of creation can lead to deeper love for God and a more humble and open appreciation of his character. Eco-physical awe and wonder can lead to theological awe and wonder and vice versa. To this end, Van Dyke writes:

> The church should provide and facilitate deliberate contact between people and nature . . . It can go further, and plan times for its members to *be in nature* and *learn from nature* by planning retreats, long or short in duration, in which nature is not merely the backdrop or setting for a gathering but the subject of it . . . Simple teachings about how to tell different kinds of trees, flowers, and grasses from each other, how to distinguish one bird's song from another's, and how to read the shape of a landscape and predict what you might find on hilltops, slopes, and depressions profoundly affect human perception of nature, and human enjoyment of it.[76]

In Romans 1:20, the Apostle Paul writes, "For since the creation of the world God's invisible qualities—his eternal power and divine nature—have been clearly seen, being understood from what has been made, so that men are without excuse." All creation everywhere is speaking of God's power and love. We Christians rightly cite this verse

---

73. See, for example, Stott, *The Birds Our Teachers*.
74. Meye, "Invitation," 31.
75. Van Dyke, *Between Heaven and Earth*, 219.
76. Ibid., 209–10.

Dusty Earthlings

as an apologetic for the witness of creation to God and the fact that nature, theistically interpreted, can help unbelievers grasp God's existence and character. But I think the verse has meaning for us Christians as well. It says that for us believers God has removed our blindness to creation's testimony to God. As Bible scholar Peter Stuhlmacher puts it, "*With [our] new orientation to life by faith in Christ, [we] Christians are now well able to recognize God in his work in Creation again.*"[77]

The third way we can join in creation's praise is by *being what we really are*. Let me explain. Psalm 148 calls nonhuman things to praise God—sea creatures, lightning, hail, trees, and wild animals. These things are not conscious and do not speak, nor do they make music or poetry. How can they praise God? Richard Bauckham answers: "All creatures bring glory to God simply by being themselves and fulfilling their God-given roles in God's creation."[78] Frogs praise God by croaking, birds by flying, fish by swimming, wolves by howling, horses by running, and so on. These and all creatures honor God by being what they are and fulfilling their roles as part of creation. This applies to us humans too. We who are redeemed in Christ can bring praise to God by *being what we are*—what God created us to be. So what *are* we? We are dusty earthlings who have the capacity to *be* humans in the image of Christ (God). Being like Jesus—having dominion over creation in the way of Jesus, showing proper humility, engaging in voluntary self-limitation, living kenotically, exercising justice with other humans and other creatures, all within the order and limits of the ecosphere—all this constitutes the way that we humans can glorify God with creation by *being what we are*.[79]

The fourth way is by caring for God's creation in our dominion over it. This, of course, is part of *being what we are* as human creatures in God's creation. This means that as caretakers and overseers, we nurture and facilitate creation's praise of God. Unfortunately, we humans have all too often interfered with creation's ability to do this.[80] To use a contemporary illustration, "'the heavens proclaim the glory of God' [Ps, 19:1] with less clarity on a smoggy day in Los Angeles than on

---

77. Stuhlmacher, "Ecological Crisis," 10.
78. Bauckham, "Joining in Creation's Praise," 47.
79. Rom 8:29; 2 Cor 3:18; Phil 3:21.
80. Endres, "Praising God," 107.

other days."⁸¹ But we can change our ways. We can praise God by reducing pollution, protecting species, preserving habitats and landscapes, and providing for other creatures the space and resources they need to praise God. Our nurturing and encouraging of other creatures in their capacity to live and bring honor to their creator glorifies God, fulfills our dominion role in a Christlike way, and concretely recognizes the inherent value of God's nonhuman creatures.

This chapter is about Christian ethics—what we ought to be and do as followers of Jesus. Praising God with his creation is something we ought to be and do. We may not be in the habit of doing this, but, if we work at it, we can learn and develop new habits. At this point, someone may say that they have to work and care for their families; they cannot be concerned with being wondering onlookers or with joining creation's praise of God. Of course, as physical beings, all of us must be concerned with our families and our livelihoods. To work, cultivate land, raise crops, produce goods, build shelters, and make our living from God's earth (within reasonable limits)—all of these activities are part of being dusty earthlings, and they all bring glory to God. There may be times when immediate needs or suffering may crowd out any notion of praise, thanksgiving, or joy. But this does not mean that praising God with creation is not a valid part of human spirituality and life. This is not only a practice; it is also an attitude, a perception, a consciousness, a way of *being* in the world. Besides, even while we work to meet our basic needs, we can enjoy creation and praise God with it. "It is distinctively human to bring praise to conscious expression in voice, but the creatures remind us that this distinctively human form of praise is worthless unless, like them, we live our whole lives to the glory of God."⁸²

Conclusion

Robert Costanza and his colleagues write in their book on ecological economics that virtue and principled action by the individual person, prior to any public, governmental, or corporate action, is the key to ecologically responsible living:

---

81. Fretheim, "Nature's Praise," 29.
82. Bauckham, *Bible and Ecology*, 79.

Dusty Earthlings

> These formal institutions, markets, governments, and voluntary organizations, though potent forces, should not cause us to overlook the most fundamental source of power in an open society, namely, the actions and values of individuals . . . Individual actions and values are the ultimate determinants of environmental quality and of the possibility for sustainability. Individual decisions about what to purchase, consume, wear, and drive, about where and how to live, what jobs to seek, how many children to have, will decide the future . . . Few would argue that government alone, relying upon current practices, can be depended upon for environmental protection.[83]

This seems like quite an invitation from a secular environmentalist to us the church of Jesus Christ to live the gospel and show the transforming power of God in our eco-physical lives. Are the gospel of Christ and the values of the kingdom of God manifested in the lives of his followers (us) such that the answers that Dr. Costanza and his colleagues are looking for are evident? As we have seen, we Christians have wondrous resources for personal and corporate transformation and eco-physical witness to Jesus Christ. Are we using them?

---

83. Costanza et al., *Introduction to Ecological Economics*, 185, 187.

8

Conclusion

Summary

BEFORE OFFERING SOME CONCLUDING thoughts, allow me to summarize. Again, my basic argument is as follows: (1) We humans are *physical beings, dusty earthlings*. (2) Since we are physical beings, we are also *eco-physical beings*, embedded in and dependent upon the earth's ecosphere. (3) As eco-physical beings, we are subject to all the principles, patterns, parameters, and limits (*ecological realities*) that govern ecological existence on this planet. (4) These realities have implications for Christian ethics—the kind of people we should be and how we should live on God's planet. Our worldview, our attitudes, and our way of life ought to reflect the reality of our eco-physicality, our creaturehood, our citizenship in God's ecological community of creation.

If adopted, this would be a new approach for us because we perceive ourselves as separate from the rest of God's creation. We are alienated from nature. This alienation issues originally from our primal rebellion against God (Gen 3), but it has been exacerbated by some philosophical, religious, cultural, scientific, and technological developments in the modern era, so that today nearly all of us who live in affluent, industrial societies think, feel, and live in isolation from the natural world around us. While many scientific developments and their resulting technologies have increased our separation from nature, scientific investigations of humans and other animals have demonstrated our solidarity with nature. Neuropsychology, ethology, primatology, anthropology, and ecology have shown that we humans are organic, physical animals, creatures among creatures, just like all the other animals of earth. We share many characteristics with them including basic biochemistry, physiology, social life, and even rudimentary culture and cognition. These commonalities affirm our eco-physical nature. But science has also identified human characteristics that distinguish us from other creatures. These include language, complex culture, self-transcendence, free choice, love, and religious life. As a result, we are God's *ethical species*, his *moral animals*, possessing capacities that justify God's charging us with the care of his earthly creation.

Looking at Scripture, we found that it teaches theocentrism—that all creation is centered on God. God is the creator, owner, and provider of all things. He has a direct relationship with his creation and his creatures. God delights in the beauty, diversity, and operation of his creation as a whole (Gen 1:31). In the acts of creation and redemption, God acts kenotically as a God of self-emptying love.

Humans are creatures, not gods. We are of the earth—integral members of creation (Gen 2:7). Only God is above and outside of creation. The biblical view of the world is holistic. God created it, provides for it, and redeems it as a whole—and us as a part of that whole. This holistic view has implications: (1) Before all else we are members of the worldwide human family whose mother and father are Adam and Eve, but at a deeper level we are, with all creatures, members of God's whole creation. (2) There is no division between spiritual and material. These are merely two aspects of the single *spiritual-physical reality of our creaturely existence*. (3) As creatures who are part of creation, our existence is characterized by contingency, dependency, and limitedness. Moreover, Jesus Christ, in his dusty incarnate being, is the paradigmatic human being—the one true image of God in his world.

The biblical idea of the image of God (Gen 1:26) applies to our whole physical-spiritual-ecological being as creatures of God. As God is relational, we are also relational. Our existence is constituted in our relationships with God, with other humans, with other creatures, and with his whole creation. We represent God to his creation, and we represent creation to God. Our religious goal is to glorify God by being conformed to Christ, the true image of God.[1]

In the dominion mandate (Gen 1:26–28), God appoints us managers of his earthly creation and gifts us with power to carry out that responsibility. But since we are part of creation, we are "insider"

---

1. Rom 8:29; 2 Cor 4:4; Col 1:15.

managers—always working from within creation. Dominion involves care and protection of God's creatures (Gen 2:15), and since we too are part of creation, dominion means managing ourselves and our own desires and inclinations. We should understand our dominion over creation as being in the way of Jesus Christ. Like Christ, who creates and gives space and freedom to us that we may flourish, we should give space and freedom to other creatures that they may flourish. Dominion in the image of God means dominion in the way of Jesus.

Ecological sin consists in our denial of our eco-physical nature, our rebellion against limits, our quest to become gods—in sum, our refusal to accept our creaturehood. God's solution for our sin is holistic redemption in Jesus Christ. We humans are redeemed and given the capacity to live in the way of Jesus. But God's redemption goes further. Since our very existence is bound up in creation, and since all of creation is valuable to God, he has redeemed all of his creation to himself, us included. Christ's great work of reconciliation brings all things back to God and provides for the final transformation and restoration of all things in the day of the Lord, when Jesus comes again.[2]

I cited the concept of *theocentrism*, the idea that all creation exists for the sake of God (Rom 11:36), glorifies God (Isa 6:3), and points to God. What makes any creature, including us, special is God's relation to it, not any inherent quality that it (we) may possess. It's about God, not us.

The science of ecology has worked out some principles, patterns, parameters, and limits that circumscribe the existence of all creatures. These include feedback loops, resource recycling, energy flow and exchange, food webs, biodiversity, ecosystem services, population dynamics, the law of entropy, the principle of limits, and the new ecology. As dusty earthlings we must learn how to live within the framework of these realities in order that we and all of God's creatures may flourish and bring glory to God.

Finally, I discussed ecological ethics, which includes at least two virtues (voluntary self-limitation and humility) and two principles (kenosis and justice) as part of what we dusty Christians need to do to align ourselves with God's design for the planet and for our lives. We also have the privilege of joining in creation's praise of God.

---

2. Matt 19:28; Acts 3:21; 1Cor 15:27–28; Eph 1:10; Rev 11:15.

"Eco-Consciousness"

As a way of healing our alienation from nature and of living in a manner that is more aligned with the ecology of the planet, I propose the practice of "ecological consciousness," or "eco-consciousness" for short. This is not some kind of "mystical awareness" of nature. On the contrary, what I have in mind is a practical, down-to-earth, day-to-day, intentional cultivation of an awareness of our eco-physical life as dusty earthlings. This would involve consciously relating all of our material existence to God and to his creation. Jesus calls us to live "all of life,"[3] including our eco-physical life, in relation to God. Biologists Gene Likens and Jerry Franklin write of "eco-thinking," that is, thinking "within the framework . . . in which all components of the ecosystem (including humans and human activities) and all known factors affecting them are considered and evaluated."[4] This is what I am talking about, except that we do it within a Christian framework—in terms of our relationship with God. In our everyday thinking, evaluating, deciding, working, and acting, we would take into account our eco-physical nature and God's ecosphere and the fact that we are part of it.

We humans are highly social beings. Awareness of our social relationships and context comes naturally to us. We are continuously aware of our relationships and interactions with other people—family, friends, coworkers, fellow church members, and so on. We do this automatically. We are innately socially conscious beings. Our social consciousness comes naturally, but ecological consciousness does not. In reality, however, we *are* ecological beings as well as social beings. So what I am suggesting is that we begin to embody that reality by training ourselves to be aware of our ecological context as well as our social context. We do not normally do this, so it would require some work. Eco-consciousness will never be as natural and automatic for us as social consciousness, but I think that with God's help we could do better than we are doing now and, as a result, live lives that are more aligned with ecological realities.

Eco-consciousness in relation to God would apply to all areas of life, including relationships, work, worship, travel, living arrangements, entertainment, consumption, waste, and so on. It means that in my

---

3. Scorgie, *Little Guide*, 26.
4. Likens and Franklin, "Ecosystem Thinking," 512.

daily life, I would see myself situated within the ecological world, in relationship with God, with other people, with other creatures, with the soil, with the oceans, and with the broader earthly creation. I would be conscious of my impact on the ecosystem, on other people, on creatures, and on God's earth. I would ask questions such as the following: Am I living in a way that makes room for other people and creatures in this world? Do I have enough? Do I have too much? How can I embody virtues of humility and self-limitation? How can I live more justly in an ecological sense? Eco-consciousness is thinking *outside* the box of culture but *inside* the box of God's earthly creation.

I am not suggesting that we become legalistic about this or that all churches should make ecological care the central focus of their ministry. What I am suggesting is that whatever we do, wherever we are, whatever our ministry, we include an eco-physical perspective and the eco-physical implications of what we are doing. If you are a teacher, plumber, contractor, administrator, laborer—whatever your occupation—you would ask yourself how to integrate ecological realities into what you do. For our churches and other Christian institutions, how can we integrate an eco-physical perspective into our existing work and ministries? How can we do evangelism ecologically? How can we do worship ecologically? How can we care for people ecologically? Many churches are involved in great ministries. They should continue that great work but also integrate eco-consciousness into those ministries.

Ecological Education: Learning about God's Creation

If we are part of the ecosystem, we need to know something about how the ecosystem works. The great John Calvin wrote, "The Lord has furnished men with the arts of deliberation and caution, that they may employ them in subservience to his providence, in the preservation of their life."[5] God's providence works through the ecosystem. In order to be deliberate and cautious in subservience to God's providence, as Calvin suggests, we need to know something about ecology. Ecological knowledge will better equip us to live wisely, "in subservience" to God's design and providence, within the limits of the system, and to act in ways that preserve and protect it (Gen 2:15) for the sake of ourselves,

---

5. Calvin, *Institutes*, 1.17.10, p. 187.

other people, our posterity, all of God's creatures—and most importantly, for the sake of God. To do this, we need ecological education.

Education is not *the* solution to the Ecological Problem, but it is *part* of the solution. As we have noted, at its origin, the Ecological Problem is spiritual. Changes in worldview, attitudes, and values are required for us to return to reality and begin living like the eco-physical beings that we are. But a part of this transformation is knowledge—acquiring the information and concepts we need in order to understand ourselves and the world in terms of eco-physical reality. This is part of the purpose of this book. After all, we can only live and act based on what we know. If we do not know about something, we cannot act on it. The only way to know about ecology (or anything) is to learn about it. Besides, learning about God's creation is fun.

Learning about God's creation is part of joining with creation in the praise of God (as I mentioned in chapter 7), and part of what it means to be *wondering onlookers* of creation in the image of God.[6] As pastor Scott Hoezee writes, "If we are to be lovers of God, we should want to understand creation and know more and more about it so that we can enjoy it even more."[7] There is a great need for ecological knowledge and literacy. The *National Environmental Report Card* produced by the National Environmental Education Foundation says,

> As environmental issues become more complex and increasingly the result of accumulated individual actions, the importance of environmental knowledge on the part of each American will increase. More will be required of both individuals and their leaders in our environmental future. Environmentally knowledgeable Americans will better understand what they as individuals can do to solve environmental problems and will be better motivated to take action. A knowledgeable public can also play a larger role in evaluating whether proposed environmental laws and regulations make sense, in determining what new policies are needed, in supporting government regulations and policies, and in claiming information that it is the public's right to know.[8]

---

6. Santmire, *Brother Earth*, 190.

7. Hoezee, *Remember Creation*, 30.

8. National Environmental Education Foundation, *National Environmental Report Card*, 2.

Unfortunately, surveys show that most of us do not know much about the creatures around us or the ecosystems in which we live: "Americans are not prepared for our environmental future. Fewer than one in nine Americans gets a passing score of 60% on knowledge of issues likely to be major problems in the next 15–25 years. Just 1 in 25 scored 70% or above in a quiz of environmental knowledge. On average, Americans answered just three multiple-choice questions right on a ten-question quiz about issues in the next century."[9] Twentysomething Christian eco-activist Ben Lowe notes, "It's astonishing just how little many of us, especially in my generation, know about what it takes to sustain our everyday demands."[10] Chapter 6 was an effort to begin to address the problem of ecological education.

Moreover, we need to learn about being disciples of Jesus within our ecological context: training in the values, principles, and virtues of the kingdom of God as Jesus taught and lived them in a way that frames them ecologically. Chapter 7 was an effort in this direction. For dusty Christians, ecological education involves learning about self-limitation and how to live humbly, kenotically, and justly, in the way of Jesus, to the glory of God. Christian ethicist Larry Rasmussen writes, "Moral formation should include a keen sense of limits and knowledge of the fact that all behavior has environmental consequences. Child-rearing (and adult-rearing!) should include, in both formal and informal education and training, the inculcation of certain moral sensitivities: awe, reverence, gratitude, and vulnerability, among others. Creation is one, it is precious, and it is finite."[11]

Individual Lifestyle and Public Policy

Ecological life is unavoidably community life. It is by its nature shared among people and creatures. It cannot be individualized or privatized. My personal ecological lifestyle impacts you, and your lifestyle impacts me. We are all, willy-nilly, ecological players, part of the local ecological community in which we live, and ultimately of the entire ecosphere. No matter what we may believe about human nature or technology or

---

9. Ibid., 2–3.
10. Lowe, *Green Revolution*, 52.
11. Rasmussen, "Creation, Church, and Christian Responsibility," 130.

Dusty Earthlings

economics or anything else, we are always affecting the ecosystem in some way—for better or for worse. Everything we do—washing our faces, brushing our teeth, eating, driving, reading, talking, walking, buying, working, traveling, writing, being born, dying—*everything* we do, we do as ecological beings. All takes place within the ecosystem, is supported by the ecosystem, and affects the ecosystem. My ecology is unavoidably shared with others. Like Noah, his family, and the animals, we are all in the same boat.

But most of us Americans are individualists, and I am no exception. I like my privacy and personal freedom. I don't like other people telling me what I can and can't do, and I don't like the government doing it either. As one who believes that we Christians should not have to be forced to do what our theology should lead us to do anyway, I have tended to believe in an inside-out, bottom-up approach to the Ecological Problem. I have thought that, at least among Christians, its solution will emerge as more and more individual Christians follow Jesus in all of life, have their hearts changed, become eco-conscious, and live humble, kenotic, just lives in the way of Jesus, voluntarily limiting their material lives, changing their individual lifestyles, shrinking their ecological footprints, living more humbly in God's ecosphere. As Matthew Sleeth says in his creation care workbook, "a changed heart = a changed life."[12]

But I have come to see that it doesn't necessarily work that way. Christians don't seem to change simply on the basis of their conversion or because of their theology. Despite our presumed receptiveness to biblical ideas and the indwelling Holy Spirit's illuminative and transformative power, we continue to squabble among ourselves about ecology (and many other issues), and, like everyone, we struggle with ignorance, greed, selfishness, complacency, and sin (Rom 7:15). So I have changed my thinking. I still believe that an individualized, inside-out, bottom-up approach is essential, but it's only half the answer. Christian or not, most of us will need laws and regulations to make us do what we ought to do and change what we ought to change. As Christian ethicist Kevin O'Brien writes, we will need not only to "think differently," but also to "demand laws that force us to act differently."[13] Social scientist Jonathan Harrington, in his study of evangelicals and their response to climate change, speaks of the "search for solutions to mobilizing public

---

12. Sleeth, *Hope for Creation*, 14, 30, 48ff.
13. O'Brien, *Ethics*, 147.

support for laws and regulations that reduce human impacts on the global climate."[14] His assumption is that laws will be required in order to change behavior—no matter what one's religious beliefs may be. I am afraid he is right. The issue of public laws and regulations for addressing ecological problems is complicated and difficult. But suffice it to say that even if we Christians actually did live out our theology in our eco-physical lives (something most of us are far from doing), we would still need laws and regulations. It seems to be the nature of human societies, especially complex societies like ours. But laws or no laws, Christians who are empowered by the Holy Spirit, who study their Bibles, who are serious about Jesus, who have some ecological knowledge, and who take their dusty "earthiness" seriously, can change their ecological behavior. It takes learning, effort, discipline, and the support of church leadership, but it is possible.

Personal, attitudinal, and behavioral change is vital, but in the final analysis, there are, I believe, only five ways to change large numbers of people, such as the church or the society as a whole. These are as follows: (1) *Cultural shifts in values and understandings*. The civil rights movement and the women's movement are examples of this. By virtue of the persistent efforts of relatively small groups of people, major shifts in our assumptions, understandings, and behavior about minorities and women have occurred and continue to evolve. (2) *Governmental laws and regulations*. While acknowledging that some bad environmental laws have been passed, over the years, much good has been accomplished through legislative action. As I just noted, this is a complex and difficult area. But despite that, it will remain a part of life as we seek to live more ecologically in the world. (3) *Social pressure*. We are powerfully influenced by our social environment. If you arrived at church and found that everyone had ridden bicycles or public transit, and you were the only person who had driven a car, you would be embarrassed and would probably ride a bike or take the bus in the future. Ecological behaviors like appropriate transportation, recycling, and living "smaller" need to become socially normative among Christians. (4) *Economic factors*. If gas were relatively more expensive, say, $100 a gallon, a lot more people would move closer to their work, drive smaller cars, walk, or ride bicycles. (5) *Ecological consequences*. Events such as fuel or water shortages, famines, fires, storms, floods, epidemics, pests, oil or sewage

---

14. Harrington, "Evangelicalism," 3.

spills, or poisonous smogs that impact us personally or impact our families and friends can change our behavior.

Generally, we Christians conform to our cultures. With some notable exceptions, down through history Christianity has often been so "contextualized" that its impact on the world has been attenuated to the vanishing point. On the other hand, Christianity has been influential in effecting profound changes such as the abolitionist movement of the nineteenth century and the civil rights movement of the twentieth. And there has been considerable ecological activism on the part of some Christians.[15] But most of us will need all of the above—Christian transformation, cultural movements, social pressure, economic incentives, laws and regulations, and perhaps even ecological consequences—in order for us to change our ways and live more ecologically. As ecologists Gene Likens and Jerry Franklin comment, "Ultimately, societal commitment is essential for successful environmental policies."[16]

Population

Christians hold a variety of opinions on human population. On one side, a few are concerned about overpopulation and limit their reproduction accordingly. On the other side, some feel that children are a "blessing from God" and have as many children as they can.[17] Probably, most Christians fall in between and use modern methods of birth control to manage their reproduction according to their personal preferences and their financial circumstances. Like most developed nations, the United States is undergoing the so-called *demographic transition*. This occurs when with industrialization, urbanization, better education, better living conditions, improved public health, decreasing death rates (especially among children), and the education and empowerment of women, the birthrate declines and population growth rate slows. This has occurred in nearly all industrial societies around the world including the United States. Today the birthrate in America is about 2.1 births per fertile woman, which is close to a level that would result in a stable population. Our population continues to grow, however, due to

---

15. See, for example, Van Dyke, *Between Heaven and Earth*, chs. 4, 6, 7, 8.
16. Likens and Franklin, "Ecosystem Thinking," 511.
17. Citing Psalm 127:3–5, the Quiverfull Movement among conservative Christians eschews birth control and advocates large families. See www.quiverfull.com/.

immigration and the lag time that occurs between a drop in birthrate and population stabilization. Despite the controversy that swirls around this issue, I will offer three comments.

First, as I have been arguing, we are limited creatures, and we live in a limited world. The Bible does contain the idea that having children is a blessing from the Lord, but to interpret Psalm 127:3–5 (and related passages) as a mandate for unlimited reproduction seems to be a narrow, wooden, and injudicious approach to Scripture. The Bible also contains other ideas.[18] Overall, it does not provide a mandate for unlimited human procreation. God commanded the animals to "be fruitful and multiply and fill the earth"—not just humans (Gen 1:22, 28). Jesus himself, as the gospels describe him, did not marry and had no children. Paul seemed to favor celibacy yet suggested that there is some range of marital and reproductive choices that are acceptable within the framework of Christian commitment and ministry (1 Cor 7:1–40). One persistent biblical theme is that we humans are not limitless creatures living in a limitless world. We are ecological creatures and we are rational creatures. We have a God-given capacity to examine our circumstances and stand back and see ourselves as living *within* the world; we can process information, look to the future, and make thoughtful choices. Therefore, in considering our reproductive lives, we can legitimately take into account the limited resources of the world, their just distribution among people, the effect that our reproduction will have on the ecosystem that supports us, and the conditions in which our offspring might have to live. These considerations, it seems to me, should lead one to consider rational, prayerful control and limitation of one's reproductive life.

Second, as I noted in chapter 7, the problem of population is tied to the issues of justice for the poor and the fair distribution of resources. Susan Power Bratton points out that in this area, our Christian track record has not been good: "The history of population regulation in Christendom, with its ghosts of female infanticide, abandoned infants, and a million Irish expiring in a single food crisis, clearly indicates that Christians should be very concerned about population issues because they are so closely tied to God's call for justice for the defenseless, the poor, and the oppressed. Overpopulation breeds conflict and injustice as resource availability falls, spurring people to wrest limited necessities

---

18. For example, Matt 19:11–12; 1 Cor 7:32–35.

from one another. The reverse is also true. Injustice breeds overpopulation and population-related problems."[19] The population problem is not merely a problem of numbers; it is a problem of justice and ecological impact per person. "If we rated population problems not by total head count but by magnitude of environmental resource consumption and environmental damage, we would quickly come to the conclusion that some of the industrial democracies have the worst population problems of the world."[20] This points the finger right back at us who live in the so-called developed nations of the world. We contribute significantly to the "population problem." We are consuming an unfair proportion of the world's resources and are being wasteful in our use of them. We Christians, who are called to love our neighbor and who should be concerned about justice, ought to consider limiting all aspects of our lives, including our reproduction, in order to address this inequity. If we combined reproductive self-restraint with voluntary self-limitation in other areas of our lives, we could have a significant positive impact on the world for ecological justice.

Third, humans are not, or ought not to be, mere animals, like rats or lemmings or snowshoe hares, which, if placed in a felicitous environment, will reproduce willy-nilly until they exceed the carrying capacity of that environment and crash. As ecologist R. V. O'Neill notes, "the combination of human population growth and increasing per-capita impact is placing irreconcilable demands on the global biotic system . . . The population *will* reach a limit . . . The population *will* be controlled."[21] How it will be controlled is up to us. Unlike other animals, we humans can decide this issue. As I have argued, we humans *are* animals, but we are special animals, endowed with capacities of rational thought, self-transcendence, moral responsibility, and religious life. Reproducing without restraint and without thoughtful consideration of our creatureliness, eco-physical limits, and the needs of others is not consistent with our human uniqueness, nor is it a Christian approach. We have the capacity to understand the world around us and its limits and to rationally control our reproduction accordingly. Self-limitation in all aspects of life, including reproduction, is a normative aspect of what it means to be a Christian human.

19. Bratton, *Six Billion*, 95.
20. Ibid., 133.
21. O'Neill, "Perspectives," 1031, 1033.

Ecology and Economics

Since we are physical beings, economics—the production, exchange, consumption, and disposal of goods and services—is basic to our lives. Jesus understood this, and he had a lot to say about it. Christian biologist Fred Van Dyke remarks that our "economic behavior is arguably the most accurate expression of national, corporate, community, and individual values."[22] No matter what we may say or believe about ourselves, how we make and spend our money and the things we own and consume are defining features of our spirituality and moral condition. I am not an economist, but since my argument is about our physical nature, and economics is directly involved with that, I want to offer two comments about economics.

First, in today's world, we think about our modern economic system and evaluate its performance in isolation from the ecosphere, which, in reality, surrounds and supports it.[23] In our modern worldview, economics and ecology have traditionally fallen into separate categories. Economic indicators such as the gross domestic product (GDP), consumer demand, consumer confidence, market prices, and so on, do not generally take into account ecological realities and limitations.[24] If we are eco-physical beings embedded in and dependent upon the earth's ecosphere, then our economic system is too. We live and function economically *within* the ecosystem. Or, in the language of business, our economy is quite literally a "wholly owned subsidiary of the environment."[25] As we noted in chapter 6, the global ecosphere provides a vast array of services to support our economic activity. These include such things as provision of breathable (and burnable) oxygen, fresh water, soil formation, fuel reserves, the absorption of wastes, and so on. If we assigned market values to these services, the total would rival the world GDP.[26] Ignoring these services is analogous to managing a household while ignoring the outside services that allow it to exist in the first place: inputs of breatheable air, utilities (water, electricity,

---

22. Van Dyke, *Conservation Biology*, 384.
23. Daly and Farley, *Ecological Economics*, 15.
24. See McNeill, *Something New*, 334–36.
25. This quote is attributed to Timothy Wirth, former Colorado senator and undersecretary of state for global affairs, and current president of the United Nations Foundation. Quoted in Houghton, *Global Warming*, 225.
26. Costanza et al., "Value of the World's Ecosystem," 253–60.

fuel, food, clothing, repairs), and the removal of wastes (sewage, trash, used items, etc.). Our economic system exists *in the world*, not in some disembodied dimension separate from the earth. Since we and our economy live within the ecosystem, our economic thinking, evaluation, management, and planning ought to consider ecological factors. Thankfully, as I shall mention shortly, a small minority of economists are beginning to do just that.

Second, in order to sustain itself, our modern economic system in its current form requires ongoing growth in terms of increasing production, consumption, material and energetic throughput, and wealth accumulation.[27] Current economic assessment defines the "health" of the economy as a "stable and high *rate of growth*."[28] Ever expanding rates of production and consumption are essential. Otherwise, the economy stagnates and begins to die. The overriding concern of businesspeople and politicians alike is to ensure immediate and continued economic growth. An ever expanding economy becomes the final goal such that all else is subsumed by it, including nonhuman creatures, ecosystems, and even people.

Obviously, since we are physical beings, an economic system is essential to provide for our needs and survival. But today, we have gone far beyond needs and survival. The imperative of economic growth has resulted not only in expensive government bailouts when growth falters, but also in a cultural mentality that seems to worship growth.[29] Evidently, the nature of the system is such that it requires ever growing consumption and the growing pursuit of wealth forever, just to keep it going. Additionally, the growth imperative seems to have resulted in a reversal of priorities. The system no longer serves people; people serve the system. To a certain extent, we humans are no longer the goal of economic activity; we are the means. No longer does private enterprise produce goods and services to serve people; people consume goods and services to serve private enterprise.[30] Human populations

---

27. Throughput is defined as "the flow of raw materials and energy from the global ecosystem, through the economy, and back to the global ecosystem as waste" (Daly and Farley, *Ecological Economics*, 6).

28. Costanza et al., *Ecological Economics*, 148.

29. McNeill, *Something New*, 335.

30. Bodley, *Anthropology and Contemporary Human Problems*, 19–22. He bases this section on a book by Leslie Sklair, *Sociology of the Global System*.

around the world are seen not as persons but as consumers, markets, or potential markets, to be exploited and controlled by generating in them the consumption of goods and services required by corporations and businesses to produce ever increasing profits and wealth and keep the system going. We overproduce, so we must overconsume. We *must* buy; we *must* eat; we *must* use; we *must* consume, *more and more*, all the time, in ever increasing quantities, at ever increasing rates, or the system will falter and begin to die. We all are afraid of this happening, so fear drives some of our consumptive behavior. In the late 1940s, the economist Victor Lebow wrote, "Our enormously productive economy . . . demands that we make consumption our way of life, that we convert the buying and selling of goods into rituals. We need things consumed, burned up, worn out, replaced and discarded at an ever increasing rate."[31] Consequently, there is a great emphasis on advertising and marketing in order to generate the needed buying and consumption. We live by the Cartesian mantra "I consume, therefore I am."[32] People have become cogs in a great global economic machine. We are caught up in a vicious circle of production and consumption.

Simple logic would say that a global economic system that depends on limitless growth but operates within a limited global ecosystem cannot survive indefinitely. At some point it will encounter limits and will either change or die. As we have already seen, there is evidence that our current rates of consumption and waste production are not sustainable long term. But modern economic thinking says that market capitalism *is*, in effect, limitless in its capacity to improve efficiency, to find new resources, to develop alternative ways of using existing resources or substituting for them, and to deal with pollution and ecological degradation. As a result, it is believed that there are *no* limits to growth. The mechanism for this is the *market*, and the means is *technology*. "Energy, resource, and pollution limits to growth . . . will be eliminated as they arise by clever development and deployment of new technology."[33] Philosopher Mark Sagoff points out that the modern economic system is dynamic. It does not simply distribute a fixed amount of resources. "In an actual market, entrepreneurs who have good ideas can increase

---

31. Paul Ekins, "The Sustainable Consumer Society: A Contradiction in Terms?" quoted in Bouma-Prediger and Walsh, *Beyond Homelessness*, 172.

32. Santmire, "Consumerism to Stewardship," 338, n.5.

33. Costanza et al., *Introduction*, 148.

the resource base by improving technology. In other words, the actual market is dynamic, not static. Rather than allocating a 'given' bundle of resources, the market through innovation leads to the creation of many new bundles of resources—such as the Internet—often unlike any of those known before."[34] Innovation and technological development driven by market demand expressed through prices determine what counts as a resource and constantly expand and change resources.[35] History (so far) supports this argument. Time and again, in the face of scarcity (limits), people have developed new technologies to exploit new resources, developed new ways of using old ones, improved efficiency, and developed new ways of doing things so that apparent limits were surpassed or simply became irrelevant. In fact, this may be occurring right now as "green" energy sources (wind, solar, geothermal, etc.) are being developed in the face of rising fuel prices, anticipated price instability, climate change, and declines in known fossil fuel reserves.[36]

So *is* economic growth limitless even though the ecosystem on which it depends is limited? Again, I am not an economist, but I suggest that the answer must be no. Here are three reasons: First, as I discussed in chapter 4, claiming that any aspect of human affairs is limitless is theologically problematic. There is only one being in existence to whom we may attribute limitlessness, and that is God. We are created beings, and as such, there is nothing about us or our affairs that is limitless. Remember the Tower of Babel? From a biblical perspective, the claim that we humans, our activities, our institutions, our artifacts, or our economy are limitless in any respect can only be judged dubious at best and idolatrous at worst. Historian J. R. McNeill suggests that economic growth has become a "state religion,"[37] in effect an idol (Gen 3:5; Exod

---

34. Sagoff, *Economy of the Earth*, 33.

35. Ibid., 34.

36. The late economist Julian Simon was a strong advocate of this view. In fact, Simon was so sure of the power of human intelligence, science, technology, and the market system that he explicitly claimed that the potential of the human economy was limitless. Although he was a materialist, he believed that humans are not like other animals. They are not subject to ecological principles but are effectively limitless in their capacity to transform the world and generate wealth (Simon, *Ultimate Resource*, 298–99).

37. McNeill, *Something New*, 335.

20:3). It seems clear to me that, from a Christian, biblical perspective, limitless economic growth is inherently problematic.[38]

Second, we must ask the question, From the viewpoint of the gospel of Jesus Christ and the kingdom of God, is limitless growth of wealth a goal that Christians can embrace? As I mentioned in chapter 7, there is considerable biblical material that criticizes the pursuit of wealth.[39] Fred Van Dyke notes that "Jesus had absolutely nothing good to say about it."[40] Our values and goals should lie elsewhere (Col 3:1–7). I don't deny that we should seek to provide for our needs through economic activity. This is necessary and good. But I do deny that we should hold up the limitless pursuit of wealth, and consequently economic growth, as a worthy goal to pursue for ourselves or for the economic system of which we are a part.

Third, we should admit that despite the theological problems I have noted, historically, technology and free market capitalism have a remarkable record of resiliency, flexibility, and innovation. Indeed, they have been stupendously successful at generating wealth. But does it follow that because it has behaved this way in the past, it will continue to behave this way into an endless future? Yes, in the past capitalism has been very productive, but this does not mean that it will be infinitely productive forever. Besides, history contains many examples where in various places at various times humans and their economic systems ran up against the limits of their ecosystems to the detriment of their economies and themselves. Jared Diamond's book *Collapse* catalogues several of these instances.[41] I would suggest that some growth in some places is possible (and needed), but limitlessness and infinite growth, everywhere and forever, is clearly neither possible nor desirable. There *are* limits to growth. The doctrines and theories of economists notwithstanding, we ordinary laypeople can see that the earth is a limited place, and "unceasing growth is not observed in nature."[42] Back in the 1960s, ecologist Paul Sears wrote, "To me, at least, it is disturbing to hear the current glib emphasis on economic 'growth' as the solution of

---

38. See also Isa 40:21–24; Jer 18:5–10; Dan 4:32, 34–37; 5:18–21; 11:36–39.

39. Prov 23:4–5; Matt 6:19–21; Mark 4:19; Luke 12:16–21, 32–34; Phil 3:10; 1 Tim 6:6–10.

40. Van Dyke, "Planetary Economies," 66–71.

41. Diamond, *Collapse*. See also Ponting, *New Green History*.

42. Costanza et al., *Introduction*, 32.

all ills. Growth, in all biological experience, is a determinate process. Out of control, say by pituitary imbalance, it becomes pathological gigantism and by no means the same thing as health. With the concept of a healthy economy there can be no quarrel, but to equate this with an ever-expanding, ever-rising spiral is to relapse into the folly of perpetual motion, long since discredited by a sane understanding of energetics."[43]

I grant that the unpredictable nature of human social and economic systems for both expansion and contraction and the dynamic nature of the earth's ecosystems make it difficult to know where the limits are. In fact, it is almost certainly the case that limits are not fixed but are constantly changing—expanding and contracting depending on a complex array of factors. But to recognize that limits are dynamic and difficult to identify does not mean that they do not exist. It just means that our capacity to identify and measure them is itself limited. The "bottom line," I suggest, is that there *are* limits and boundaries to economic growth and to all forms of human activity, but we have trouble knowing exactly what they are. We are like a dimly seeing person enclosed in a room. We know there are walls, but we cannot be sure exactly where they are until we bump into them—and the walls are moving in and out all the time, making the room bigger or smaller, partly related to what we do. But the fact that we cannot see the walls, or see them only dimly, does not mean they are not there. In the end, we have to accept that the economic system lives within the earth's ecosphere, and the earth's ecosphere is finite and limited. So thinking, studying, and talking about ecological limits and seeking to live within those limits as the limited creatures we are, it seems to me, are wise things to do (Luke 14:28–33).

Ecological Economics

A small minority of economists today are doing just that. They recognize that the economy is an integral part of the ecosystem and are seeking to integrate economics and ecology. These "ecological economists" are developing new ways of thinking about economics and doing business so that ecological factors and limits are taken into account. Herman Daly, Robert Costanza, Richard Norgaard, and Paul Hawken are just a few of those who are working in this area. Robert Costanza and

---

43. Sears, "Ecology," 13.

his colleagues summarize the basic principles of ecological economics. I paraphrase them here:

1. The earth's global ecosphere is limited and the human economy is a subsystem of the earth's system.

2. There are limits on biophysical throughput of resources from the ecosystem, through the economy, and back to the ecosystem.

3. The earth and all its subsystems (ecosystems and all human social and economic systems) contain a "large and irreducible" uncertainty. Furthermore, some processes (such as burning fossil fuel or mining resources) are irreversible.

4. Because of these limitations, our economic activity and management should function within these constraints in order to achieve sustainability for both humans and nonhumans, and we should be precautionary (humble and careful), proactive rather than reactive, and adaptive (able to change as conditions change or when something doesn't go right).[44]

This approach seems to me to be more aligned with a biblical view than conventional economics. It is more holistically integrated; it is about people more than about money; it embodies the virtues of humility and caution; and it is better aligned with ecological realities.

Voluntary self-limitation and kenosis in our economic behavior would, perhaps, be a part of moving toward a more just and sustainable economy. Some may argue that exercising self-control and moderation in our economic life would cause more damage since the system depends on consumption. As theologian James Nash says, the frugality that would result from such behavior is an economically "subversive virtue because it is a severe threat to an economic system that depends on compulsive consumption to keep the system going and growing."[45] If only a few people did this, it would have no impact, but if a lot of people did it, the drop in consumer demand would result in falling sales, unemployment, and recession. But we should remember two things. First, the market system is flexible and resilient. I think that our current form of "consumer-growth-capitalism" is not the only form of capitalism. Capitalism can exist in many ways, shapes, and sizes.

44. Costanza et al., *Introduction*, 79–80.
45. Nash, "Toward the Ecological Reformation of Christianity," 12.

Dusty Earthlings

A shift toward eco-consciousness, self-restraint, and more reasonable consumption patterns would result, I think, in the system changing to accommodate that. We would still have capitalism and markets, but they would simply be based on better (more Christian?) values such as humility, self-control, moderation, justice, and kenosis. Fred Van Dyke asks, "How can we better choose what we shall value, instead of treating our appetites, wants, and desires as givens that must be satisfied regardless of environmental cost? And how can we restructure the human economic enterprise so that it not only ceases to degrade the world, but makes the human presence an agent of biodiversity conservation?"[46] In short, we need to "make economic activity the reflection of value rather than the determinant of it."[47] Here is the crux of the matter. Our values should change so that we value the things that really are valuable. And here is where Christianity comes in—valuing what God values, and conforming our economic values to that. We need *Christian* ecological economists who can help us move toward this goal.

Second, the economy as it currently exists is not sustainable anyway. Sooner or later, it will have to change and conform to ecological realities. Why not begin now, before ecological constraints (or disaster) or government regulations force us to change? This is an area in which we Christians could provide leadership as we live Christ in all of life, including our economic life.[48]

Loving God as Dusty Earthlings: Beyond "Green" Living

At this point in books on ecology, authors often give a list of practical things we can do in order to be "green." But rather than give you a list myself, I will refer you to some excellent resources. A very extensive coverage of "green" living actions and practices is Nancy Sleeth's *Go Green, $ave Green*. She not only offers a multitude of ideas, she provides additional resources where you can learn the how and why of various lifestyle changes and actions. I recommend this book. Her daughter, Emma Sleeth, has written a similar book oriented toward students and young adults: *It's Not Easy Being Green*. Joanne Poyourow leads the Legacy Movement in the Los Angeles area. She and her colleagues

---

46. Van Dyke, *Conservation Biology*, 411.
47. Ibid.
48. Scorgie, *Little Guide*, 26.

are seeking to transform their lifestyles to be more aligned with eco-physical reality. She has much to teach us, and we Christians, especially those of us who live in urban settings, can learn from her.[49] There are many other sources as well, especially online. But I suggest caution here. Commercial interests may exaggerate or distort claims regarding various products or even render them downright false. "Greenness" is fashionable in many parts of the culture these days. Marketers recognize this, so caution and critical thought are in order. All that claims to be green is not green—*caveat emptor* (buyer beware).

Let me add two additional suggestions. First, as you investigate ideas and practices, try to differentiate between the philosophy (or religion) of the source you are reading and the action(s) they are recommending. The action(s) may be good while the beliefs are not.[50] But remember that as Christians we have our own reasons for doing (or not doing) things. We do not have to accept the philosophy or religion behind an action in order to accept the action itself. Also, keep in mind that this may provide opportunities for discussion of religious beliefs and bearing witness for Christ when people ask you why you are living as you do or doing what you are doing (or not doing). Second, I encourage you to read, listen, and talk to people. Become a student of ecology and environmental care. Read Christian books on ecology and creation care. There are many good ones out there. Over time, you will slowly gain knowledge and perspective.

While I support the measures suggested by Nancy Sleeth and others that will help all of us move toward greener living, I suggest that ultimately we will need to move beyond this. As Dave Bookless writes, "we cannot solve [the Ecological Problem] simply by better technology and a few hard political choices. It goes right to the heart of who we are. We need to rethink not just how we treat the planet and its creatures, but who on earth we think we are as human beings."[51] We will need to reconceive and transform how we think of ourselves and our lives within the world. But how we understand ourselves is part of our broader worldview of how we understand God and his world. In other words, the answer to the Ecological Problem is a matter of our vision

---

49. See http://legacyla.net/.

50. For example, see Javna et al., *50 Simple Things*. The book is filled with facts and good ideas, but its philosophical viewpoint is problematic. Despite this, I recommend it.

51. Bookless, *Planetwise*, 12.

Dusty Earthlings

(or worldview): how we see God, ourselves, and the creation. Our vision—our worldview—is determinative for how we live and act in the world.[52] It is this broader worldview that requires changing.

This *change* is immense. Currently, "we are not even remotely managing the richness of our world in a way that is sustainable."[53] During the twenty-first century, in order for the nine to eleven billion (or however many it turns out to be) of us to live together on this planet along with God's nonhuman creatures (if they are to survive), the massive transformation required in our attitudes, values, goals, and lifestyles is humbling, to say the least. Just as the Ecological Problem is immense and pervasive, so also is its solution. We will, sooner or later, have to radically change the way we understand God, ourselves, our world, and our place within it.

The implications of this change are profound. It means that in *all* aspects of life, we would be conscious of our eco-physical being (eco-consciousness), our place within the ecosphere, and our responsibility to God for it. We would think, feel, and live as God's creatures, members of his ecosphere, instead of ignoring this fact as we do now. Wherever we lived, whatever we did, we would understand our life, being, and actions as physical, organic, earthy, and ecological. In all things we would be mindful of our ecological dependencies and relationships in which we are embedded. We humans are active, creative beings, and that is good. The changing nature of physical existence and the inherent dynamics of the ecosphere mean that the job of being eco-physical in all aspects of life will never be finished and will require our constant attention and creativity.

It may be that this change has already begun, as there has been some movement in these areas in both the broader culture and within Christianity. But we have a long way to go. Some aspects of this change involve transcending our inherent inclinations to see the world in terms of ourselves and to live strictly in terms of economic self-interest. If we are going to wend our way through this century and the ones that follow (if the Lord tarries), we are going to have to develop a much broader, fuller, more physical, more ecological, more holistic understanding of our existence, our spirituality, and our place within God's creation.

---

52. Hauerwas, *Vision and Virtue*, 45–46.
53. Raven, "Foreword," viii.

Conclusion

A Look to the Future

Many of us baby boomers remember *The Jetsons* cartoon show from the 1960s. The program depicted a futuristic world with levitating cars, robot housemaids, instant food, and so on. Interestingly, everything was suspended in midair—separated from the earth (a kind of Platonic cartoon heaven). Perhaps the cartoon represented a kind of subliminal fulfillment of the modern project—complete separation from and mastery of earth and its constraints. Although it was only a fantasy, I think it carries a lesson for us. Such a world as the Jetsons'—of humans living in a completely synthetic, mechanized, technologized world separated from the earth—will never happen. We are earthlings. We are attached to the earth. Our future will not be somewhere up in the air or someplace other than the earth; it will be here among the rocks, plants, animals, lands, lakes, rivers, and oceans of God's good planet. The absolute conquest (obliteration, elimination, killing?) of God's nature and our separation from it envisioned by such fantasies as *The Jetsons* and similar futuristic visions will never happen.

Our modern devotion to science and technology may lead us to believe that we can remake the world (and ourselves) to our liking, but this is an illusion. God has made the world (and us) as it is. A livable and sustainable future for ourselves, for our descendants, and for God's creation will require that we take this to heart. Some technology and manipulation of nature within the framework of ecological realities is necessary and good, but we should give up our quest for final separation from the earth and the absolute subjugation of God's creation. Our place is here on earth within the organic fabric of God's world. Our future will be earthly and ecological, or it will not be at all.[54]

For us Christians our ultimate futuristic vision is the blessed hope of Christ's return, the resurrection of the dead, the judgment, and the new heavens and the new earth—the renewal of all things.[55] God will transform and heal all of his creation and us with it. For us humans, it is our nature to live "*in* our mother's womb, *in* a family, *in* a community, *in* a society."[56] And the list goes on—*in* an ecosystem, *in* a biosphere, *in*

---

54. This is a modification of a quote attributed to Thomas Cahill, "The twenty-first century will be spiritual or it will not be," quoted in Scorgie, *Little Guide*, 27.

55. Isa 65:17; 66:22; Mal 4:5; Matt 19:28; Acts 3:21; 1 Cor 5:5; 1 Thess 4:16–17; 5:2; 2 Pet 3:10–13; Rev 21:1–4.

56. Conradie, *Ecological Christian Anthropology*, 147.

a planet, *in* a solar system, *in* a galaxy, *in* a universe. Thus it is now, and thus it will be then. We are and ever shall be creatures in and of creation. In the words of James Dunn: "The recognition of humankind as a corporeal species leads directly to the confident hope that God will provide an appropriate environment for embodiment in the age to come."[57] When the Holy City descends, and heaven and earth are joined, we, in our resurrected state, will be bone and sinews, just like the resurrected Jesus (Luke 24:39). We will remain forever dusty earthlings in and of the earth our home.

We Christians are indeed blessed with this great hope of Christ's return and the final redemption of all things, but we must not forget that this hope has implications for how we live here and now. The purpose of eschatology is ethics—moral life and action—"living in the promise of the future of God in Christ."[58] As the Apostle Peter says, "So then, dear friends, since you are looking forward to this, make every effort to be found spotless, blameless and at peace with him" (2 Pet 3:14). God loves us and he loves his earth. It is not our job to bring in the kingdom of God or to save the planet. Only God can do that. But it *is* our job to live in a way that points toward the kingdom of God and his healing of the planet when Christ returns to reign supreme in his world (Rev 11:15).

---

57. Dunn, *Theology of Paul the Apostle*, 101.
58. Smith, "Theology of Hope," 533.

Glossary

abiotic: The nonliving components of an ecosystem, such as air, water, light, minerals, nutrients, mountains, hills, rocks, streams, lakes, seas.[1]

anthropocentrism, anthropocentric: The belief that everything is centered on humans, that humans are the only beings with inherent moral value, and that all of creation exists for the purpose of serving humans; cf. biocentrism, theocentrism.

anthropology: As used in this book, a doctrine (belief) about human nature—what humans are and how they are put together or constituted; cf. human constitution.

autotroph: An organism that can live and synthesize its own organic molecules using inorganic molecules such as carbon dioxide from the air and water and nitrates from the soil (photosynthesis), e.g., green plants, blue-green algae, photosynthetic bacteria.[2]

biocentrism: The belief that inherent value lies in the ecosystem or biotic community as a whole. Biocentrism often sees all living things to be of equal value;[3] cf. anthropocentrism, theocentrism.

biodiversity: The "entire array of earth's biological variety, contained in genes, populations, communities, and ecosystems."[4]

biosphere: All living things on earth considered together as a whole.

biota: The living things that exist in a given region or ecosystem; cf. abiotic.

biotic: Pertaining to living things within a region or an ecosystem or on earth as a whole: the plants, animals, and microscopic organisms; cf. abiotic.

---

1. Odum and Barrett, *Fundamentals*, 511.
2. Molles, *Ecology*, 569.
3. See Taylor, "Ethics of Respect for Nature," 74–84.
4. Van Dyke, *Conservation Biology*, 444.

Glossary

climax community: After a disturbance such as a fire or a volcanic eruption has wiped out nearly all living things in an area, organisms move back into that region through a process of *succession* in which different communities of plants and animals occur one after the other over time. Finally a "climax community" is reached that remains relatively stable as long as it is not disturbed. More recent evidence has shown that climax communities are not static but "constantly change in response to disturbances, environmental change, and their own internal dynamics";[5] cf. disturbance, succession.

creation: All that exists other than God.

culture: "The structured customs and underlying worldview assumptions [by] which people govern their lives."[6] Culture is dynamic and is formed through history, human thought and creativity, social interaction, and the ecological environment.

demographic transition: As societies become industrialized and urbanized, and as women and all citizens are educated and become empowered, birthrates decline and population growth slows.

dichotomism (also called body-soul dualism): A view of human constitution (anthropology) that sees humans as consisting of two parts: a physical body and a nonphysical or "spiritual" soul or mind; cf. dualism, monism.

disequilibrium: Living systems, such as humans and ecosystems, are not in equilibrium internally or externally with their surroundings. There are differentials or disequilibriums of order, temperature, chemical concentrations, and so on, between the system and its surroundings and within the system itself. Living systems expend energy to maintain these disequilibriums. For example, our bodies maintain a temperature of 37° C (98° F), which is rarely the same as the ambient temperature. Similarly, ecosystems are not uniform fields of temperature, chemical concentrations, and so on, but are complex, lumpy, and uneven fields of disequilibrium. A rabbit, for example, living in a meadow represents a focus of concentrated order in contrast to its surroundings that also manifest order but to a lesser degree.

---

5. Molles, *Ecology*, 456.
6. Kraft, "Culture," 385.

disturbance: Any event or process that upsets or changes an ecosystem and destroys large numbers of organisms. Examples include fire, storms (hurricanes, tornadoes, blizzards, etc.), droughts, floods, volcanic eruptions, disease epidemics, invasions of predators, parasites, and pests. Humans also cause disturbances, such as wars, logging, draining of wetlands, burning, building cities, dams, croplands, and so on.

diversity: See biodiversity.

dualism: The belief that reality or some part of it is divided into two parts that are different in substance, purpose, or moral nature. Some types of dualism important for this book are: (1) *body-soul dualism*. Belief that human beings consists of two parts: a nonmaterial soul or mind and a material body. There are many variations of this;[7] cf. dichotomism. (2) *God-creation dualism*. Belief that the fundamental division of existence is between God and creation, each regarded as a different kind of being. God is creator (uncreated existence); all else is created (created existence).[8] (3) *human-nature dualism*. Belief that humans are distinctly different from the rest of creation (or nature), each falling into different categories of being. The classic example of this is Descartes' belief that humans are immaterial minds or souls inhabiting material, machine-like bodies and that all other organisms and nature as a whole consist of material machines. (4) *spirit-material dualism*. Belief that all of existence is divided into two parts: an immaterial, spiritual world and a material or physical world.

Ecological Problem: the array of difficulties associated with our troubled relationship with the rest of creation: pollution, species loss, land degradation, deforestation, overexploitation of resources, overpopulation, and so on.

ecological realities: As I am using the term in this book, the principles, patterns, parameters, and limits of the ecosystems in which we live here on God's earth.

ecology, ecological: In this book, these words have two meanings: (1) the science of ecology that studies living things and their relationships with other living things (biota) and the nonliving elements of an ecosystem (abiotic factors); and (2) to replace the word *environment*.

---

7. See Grenz, *Theology*, 158–64.
8. See Erickson, *Christian Theology*, 397, 400–4.

Glossary

> Where others would say "the environment" or "environmental," I say "the ecology" or "ecological." For instance, I say that we should be mindful of our "local ecology" (of the local ecosystem in which we live) rather than saying we should be mindful of "our local environment."

ecosphere: The ecosystem of the entire earth including all the smaller ecosystems of living and nonliving things existing and interacting within the complexes of populations, relationships, exchanges, feedbacks, and cycles, within the thin layer of water, air, and the earth's crust around the surface of this planet. The ecosphere encompasses all humans, their artifacts, relationships, social and economic systems, and activities; cf. biosphere.

ecosystem: The ensemble of populations of various organisms living in complex relationships, interacting with one another and with nonliving elements (air, water, soil, rocks, topography, climate) within a prescribed physical area.

Enlightenment: The period of history in Western civilization from about the seventeenth century to the end of the eighteenth century when several cultural and intellectual changes took place, including a turn away from the Bible, the church, and ancient authorities to the human self and reason as sources of truth; the rise of science and technology; the emergence of a Newtonian mechanistic, atomistic view of the world; the emergence of strong individualism; and the establishment of personal freedom and autonomy as firm cultural values.

entropy, law of entropy, second law of thermodynamics: Entropy is disorder or the amount of disorder that exists in a system. It is also understood as a measure of the amount of unavailable energy in a system— energy that cannot be used to do work. The law of entropy or second law of thermodynamics says that in any closed system, total entropy (disorder) always increases. Thus, if in a given location within a system, things become more ordered (decreased entropy), other places must become less ordered (increased entropy) such that overall there is a net increase in entropy.

environment, environmental: The physical surroundings in which an organism or group of organisms live—that is, all the factors and conditions that make up the locality in which it lives. For example, a cottonwood tree's *environment* may be a warm climate along the edge of a stream or river. In this book, when I use the word *environment*, it will be in this sense of the physical surroundings of an organism; cf. ecology, ecological.

eschatology: The last things in reference to either the death of individual humans or the end of the world. *Personal eschatology* refers to what happens when each of us as individuals die. *General eschatology* refers to "the ultimate climax or end of history wherein Christ returns to earth to establish his eternal kingdom of righteousness and justice"[9] on earth among all nations and among his creatures.

ethics: The study of morality—right and wrong in human life and conduct. There are many approaches to ethics, but for the most part, in this book I use only two: (1) *virtue-based ethics*: the notion that moral rightness is manifested in being a particular kind of person who exhibits virtues such as self-control and humility; (2) *principle-based ethics*: the idea that moral rightness involves following certain principles such as kenosis and justice.

ethology: The study of animal behavior, especially in natural settings (as opposed to the laboratory). Ethologists also study human behavior and compare it to that of other animals.

eutrophication: The process whereby bodies of water (lakes, ponds, streams, rivers, estuaries, etc.) are enriched with nutrients (typically phosphates and nitrates) due to polluted runoff from farms, factories, and cities, resulting in increased primary production.[10] Nutrients in the runoff encourage overgrowth of plants, usually algae, which in turn reduce the oxygen levels in the water, and that in turn reduces populations of animals that depend on oxygen such as fish and invertebrates (e.g., crayfish and clams). The end result is reduction in species diversity and in the overall productivity of the ecosystem. For example, a lake or stream may become foul-smelling, covered with algal scum, and devoid of fish.

fertility: The average number of offspring produced by a given species population in a given period of time. For example, in human populations, fertility is typically expressed in number of live births per one

---

9. Grenz et al., *Pocket Dictionary*, 46.
10. Odum and Barrett, *Fundamentals*, 517.

thousand persons per year.

habitat: The place(s) where conditions are adequate for an animal or plant to live.

holism: A philosophical and theoretical approach that sees the whole as greater than the sum of its parts. "Connections and relations between parts must be considered, not just the parts themselves."[11] For example, holism says that in order to understand how an ecosystem works, we must consider not just the biochemistry, life, and behavior of the individual animals and plants, but also the relationships and interactions among them and the behavior of the system as a whole; cf. reductionism.

human constitution: Doctrine (beliefs) about human nature; cf. anthropology.

landscape: An area of land covered by a variegated array of patches and habitats.[12]

material: See physical.

monism: As used in this book, the doctrine of human constitution or anthropology that sees humans as a single, unified entity that is physical.

morality: The area of human life that concerns right and wrong behavior, virtues and vices.

nature: The created world of living things, nonliving things, and ecosystems.

neuroscience: The study of the brain, how it works, and how it is related to mind, consciousness, thought, emotion, experience, and behavior.

niche: The "functional role of a species in a biotic community or ecosystem."[13] The niche can be thought of as the job of a species in the ecosystem—the place where it lives and how it makes its living.[14]

nonlinear, nonlinearity: Multidimensional or complex such that no single linear sequence of relationships or of cause and effect can fully explain the behavior of something. In other words, there are multiple, interactive, dynamic factors involved in an ever changing web of

---

11. Keller and Golley, "Introduction," 2.
12. Turner, "Landscape Ecology," 78–80.
13. Odum and Barrett, *Fundamentals*, 525.
14. Molles, *Ecology*, 302.

relationships that are causing what we are observing. An ecosystem or any given part of an ecosystem (such as a species population) exists within a multidimensional matrix of elements and relationships that all behave and interact in complex and dynamic (nonlinear) ways.

ontology, ontological: A philosophical term that refers to the being or existence of something. To speak of ontology is to speak of what something actually is in itself. For example, we say that God's being is relational, meaning that God's basic existence involves relationships within the Godhead (Trinity) and with creation. Similarly, I am arguing in this book that humans are ontologically physical beings, meaning that, although nonphysical components or aspects such as a soul may be involved, our fundamental being is physical.

patch: An area of an ecosystem or landscape that is relatively homogeneous and is different from its immediate surroundings. Examples are a stand of forest surrounded by cropland or a meadow surrounded by forest.

physical: Consisting of matter and/or energy; having a material existence that humans can detect using their senses. I am arguing in this book that all aspects of human life are physical, including mental, moral, and spiritual life.

population dynamics: How biological populations change, grow, contract, or remain stable as they interact within themselves, with their ecosytems, and other populations.

primary producer: See autotroph.

primatology: The study of primates, which includes monkeys, baboons, the great apes (chimpanzees, bonobos, orangutans, gorillas), and humans. Primatologists may compare and contrast the ecology, life, and behavior of various primates.

reductionism: A philosophical and theoretical approach that sees the form, function, and behavior of any entity as fully explainable in terms of its parts. That is, the whole can be explained in terms of its parts. For example, a reductionist might argue that humans consist of nothing but chemicals and energy and that ultimately all of human behavior will be fully explained by the behavior of those chemical and energetic components; cf. holism.

second law of thermodynamics: See entropy.

Glossary

stochastic: random, variable, probabilistic. Examples of stochastic events include accidents, fires, floods, storms, disease outbreaks, and so on.

succession: An orderly sequence of changes in an ecosystem whereby one community of plants and animals replaces another.[15] Each successive community changes the environment (light availability, soil moisture and composition, food webs, etc.), making way for the succeeding community that replaces it. For example, after a fire burns a forest in western North America, a succession of communities may then appear in its place: first, grasses, small shrubs, and wildflowers with rabbits, rodents, reptiles, and certain bird species; second, larger shrubs, bushes, deer, wildcats, coyotes; third, fast-growing pines and then broadleaf trees, bears, and elk, producing an open woodland; finally firs, hemlocks, rhododendrons may form a closed forest.

theocentrism: The doctrine that all things, including humans, are centered on God and exist for the purpose of serving and glorifying God, who is the source of all moral value; cf. anthropocentrism, biocentrism.

theology: Study and reflection on the doctrines and beliefs of the Christian faith and how they relate to life and the world in which we live.[16]

wilderness: Traditionally, a region where human impact has been absent or minimal. More recent research has shown that humans have been a part of ecosystems for millennia, and wildernesses previously thought to be untouched by humans have actually been substantially influenced by them.

worldview: "The culturally structured set of assumptions (including values and commitments/allegiances) underlying how a people perceive and respond to reality."[17]

---

15. Langston, "People and Nature," 50.
16. Scorgie, "Wonder," 19.
17. Kraft, "Culture," 385.

Bibliography

Achtemeier, Elizabeth. *Nature, God, and Pulpit*. Grand Rapids: Eerdmans, 1992.
Allen, Diogenes. *Philosophy for Understanding Theology*. Atlanta: John Knox, 1985.
Anderson, Bernhard W. "The Earth Is the Lord's: An Essay on the Biblical Doctrine of Creation." *Interpretation* 9/1 (1955) 3–20.
Augustine, Saint. *Confessions*. Translated by Henry Chadwick. Oxford: Oxford University Press, 1991.
Ayres, Robert U. "The Second Law, the Fourth Law, Recyling and Limts of Growth." *Ecological Economics* 29 (1999) 473–83.
Bacon, Francis. *The Great Instauration: Novum Organum*. Whitefish, MT: Kessinger, n.d.
Baer, Richard A., Jr. "Land Misuse: A Theological Concern." *Christian Century* 83/41 (October 12, 1966) 1239–41.
Bauckham, Richard J. *The Bible and Ecology: Rediscovering the Community of Creation*. Waco, TX: Baylor University Press, 2010.
———. "First Steps to a Theology of Nature." *Evangelical Quarterly* 58 (1986) 229–44.
———. *God and the Crisis of Freedom: Biblical and Contemporary Perspectives*. Louisville: Westminster John Knox, 2002.
———. "Jesus and the Wild Animals (Mark 1:13): A Christological Image for an Ecological Age." In *Jesus of Nazareth: Lord and Christ: Essays on the Historical Jesus and New Testament Christology*, edited by Joel B. Green and Max Turner, 3–21. Grand Rapids: Eerdmans, 1994.
———. "Jesus, God and Nature in the Gospels." In *Creation in Crisis: Christian Perspectives on Sustainability*, edited by Robert S. White, 209–24. London: SPCK, 2009.
———. "Joining Creation's Praise of God." *Ecotheology* 7/1 (2002) 45–59.
———. *Living with Other Creatures: Green Exegesis and Theology*. Waco, TX: Baylor University Press, 2011.
Bauer, Walter, et al. *Greek-English Lexicon of the New Testament and Other Early Christian Literature* (BDAG). 3rd ed. Chicago: University of Chicago Press, 2000.
Beisner, E. Calvin. *Where Garden Meets Wilderness: Evangelical Entry into the Environmental Debate*. Grand Rapids: Eerdmans, 1997.
Belovsky, Gary E., et al. "Ten Suggestions to Strengthen the Science of Ecology." *Bioscience* 54/4 (2004) 345–51.
Berkouwer, G. C. *The Return of Christ*. Grand Rapids: Eerdmans, 1972.
Bernstein, Brock B. "Ecology and Economics: Complex Systems in Changing Environments." *Annual Review of Ecology and Systematics* 12 (1981) 309–30.
Berry, R. J. "Creation and the Environment." *Science and Christian Belief* 7/1 (1995) 21–43.

Bibliography

———, editor. *Environmental Stewardship: Critical Perspectives—Past and Present.* New York: T. & T. Clark, 2006.

———. "One Lord, One World: The Evangelism of Environmental Care." In *God's Stewards: The Role of Christians in Creation Care*, edited by Don Brandt, 17–29. Monrovia, CA: World Vision, 2002.

———, and Malcolm Jeeves. "The Nature of Human Nature." *Science and Christian Belief* 20 (2008) 3–47.

Berry, Thomas. *The Dream of the Earth.* San Francisco: Sierra Club, 1988.

Berry, Wendell. *The Art of the Commonplace: The Agrarian Essays of Wendell Berry.* Edited by Norman Wirzba. Berkeley: Counterpoint, 2002.

———. *The Unsettling of America: Culture and Agriculture.* San Francisco: Sierra Club, 1977.

Block, Daniel I. "To Serve and to Keep: Toward a Biblical Understanding of Humanity's Responsibility in the Face of the Biodiversity Crisis." In *Keeping God's Earth: The Global Environment in Biblical Perspective*, edited by Noah J. Toly and Daniel I. Block, 116–40. Downers Grove, IL: InterVarsity, 2010.

Bodley, John H. *Anthropology and Contemporary Human Problems.* 5th ed. Lanham, MD: AltaMira, 2008.

Bonhoeffer, Dietrich. *Ethics.* New York: Macmillan, 1955.

Book of Common Prayer. New York: Oxford University Press, 1990.

Bookless, Dave. *Planetwise: Dare to Care for God's World.* Nottingham, UK: InterVarsity, 2008.

Bouma-Prediger, Steven. "Creation Care and Character: The Nature and Necessity of Ecological Virtues." *Perspectives on Science and the Christian Faith* 50/1 (1998) 6–21. Online: http://www.asa3.org/ASA/PSCF/1998/PSCF3-98Bouma-Prediger.html.

———. *For the Beauty of the Earth: A Christian Vision for Creation Care.* Grand Rapids: Baker, 2001.

———. "Living on the Land, Living a Christian Land Ethic." *Creation Care* 22 (2003) 4–5.

———, and Brian J. Walsh. *Beyond Homelessness: Christian Faith in a Culture of Displacement.* Grand Rapids: Eerdmans, 2008.

Bratton, Susan Power. *Christianity, Wilderness, and Wildlife: The Original Desert Solitaire.* Scranton: University of Scranton Press, 1993.

———. *Six Billion and More: Human Population Regulation and Christian Ethics.* Louisville: Westminster John Knox, 1992.

Brueggemann, Walter. *Genesis: A Bible Commentary for Teaching and Preaching.* Atlanta: John Knox, 1982.

———. *The Land: Place as Gift, Promise, and Challenge in Biblical Faith.* 2nd ed. Minneapolis: Fortress, 2002.

Burns, Jeffrey M., and Russell H. Swerdlow. "Right Orbitofrontal Tumor with Pedophilia Symptom and Constructional Apraxia Sign." *Archives of Neurology* 60/3 (2003) 437–40.

Callicott, J. Baird. "Multicultural Environmental Ethics." *Daedalus* 130/4 (2001) 77–97.

Calvin, John. *Commentaries on the First Book of Moses Called Genesis.* Translated by John King. Grand Rapids: Baker, 1979.

———. *Institutes of the Christian Religion.* Translated by Henry Beveridge. Grand Rapids: Eerdmans, 1989.

Bibliography

Campbell, Neil A., and Jane B. Reece. *Biology*. 7th ed. San Francisco: Benjamin Cummings, 2005.

Carson, Rachel. *Silent Spring*. 40th anniv. ed. Boston: Houghton Mifflin, 2002.

Center for Reformed Theology and Apologetics. "The Westminster Shorter Catechism." Online: http://www.reformed.org/documents/index.html?mainframe=http://www.reformed.org/documents/WSC_frames.html.

Clark, Anna M. *Green, American Style: Becoming Earth Friendly and Reaping the Benefits*. Grand Rapids: Baker, 2010.

Clifford, Anne M. "From Ecological Lament to a Sustainable *Oikos*." In *God's Stewards: The Role of Christians in Creation Care*, edited by Don Brandt, 51–63. Monrovia, CA: World Vision, 2002.

Cohen, Joel E. *How Many People Can the Earth Support?* New York: Norton, 1995.

———. "Population Growth and the Earth's Human Carrying Capacity." In *Readings in Ecology*, edited by Stanley I. Dodson et al., 349–61. Oxford: Oxford University Press, 1999.

Conradie, Ernst M. *An Ecological Christian Anthropology: At Home on Earth?* Burlington, VT: Ashgate, 2005.

Costanza, Robert, et al. *An Introduction to Ecological Economics*. Boca Raton, FL: St. Lucie, 1997.

———, et al. "The Value of the World's Ecosystem Services and Natural Capital." *Nature* 387/6630 (May 15, 1997) 253–60.

Cronon, William, editor. *Uncommon Ground: Rethinking the Human Place in Nature*. New York: Norton, 1995.

Daly, Herman E., and Joshua Farley. *Ecological Economics: Principles and Applications*. Washington, DC: Island, 2004.

Damasio, Antonio. *Descartes' Error: Emotion, Reason, and the Human Brain*. New York: Penguin, 1994.

Davis, Ellen F. *Scripture, Culture, and Agriculture: An Agrarian Reading of the Bible*. Cambridge: Cambridge University Press, 2009.

De Waal, Frans. *Bonobo: The Forgotten Ape*. Berkeley: University of California Press, 1997.

———. *Good Natured: The Origins of Right and Wrong in Human and Other Animals*. Cambridge: Harvard University Press, 1996.

Descartes, René. *Discourse on Method and Meditations on First Philosophy*. 4th ed. Translated by Donald A. Cress. Indianapolis: Hackett, 1998.

DeWitt, Calvin B. "Behemoth and Batrachians in the Eye of God: Responsibility to Other Kinds in Biblical Perspective." In *Christianity and Ecology: Seeking the Well-Being of Earth and Humans*, edited by Dieter T. Hessel and Rosemary Radford Ruether, 291–316. Cambridge: Harvard University Press, 2000.

———. *Earthwise: A Biblical Response to Environmental Issues*. Grand Rapids: CRC, 1994.

———. "Eco-Myths: Don't Believe Everything You Hear about the Church and the Environmental Crisis: It's Not Biblical to Be Green." *Christianity Today* 38/4 (April 4, 1994) 27–31.

———. "The Scientist and the Shepherd: The Emergence of Evangelical Environmentalism." In *The Oxford Handbook of Religion and Ecology*, edited by Roger S. Gottlieb, 568–87. Oxford: Oxford University Press, 2006.

Bibliography

Diamond, Jared. *Collapse: How Societies Choose to Fail or Succeed.* New York: Penguin, 2005.

———. *Guns, Germs, and Steel: The Fates of Human Societies.* New York: Norton, 1997.

Dorn, Harold F. "World Population Growth: An International Dilemma." *Science* 135/3500 (January 26, 1962) 283–90.

Douglas, David. *Wilderness Sojourn: Notes in the Desert Silence.* San Francisco: Harper & Row, 1989.

Dubos, René. *A God Within.* New York: Scribner's, 1972.

Dunn, James D. G. *The Theology of Paul the Apostle.* Grand Rapids: Eerdmans, 1998.

Dyrness, William. *The Earth Is God's: A Theology of American Culture.* Maryknoll, NY: Orbis, 1997.

———. "Stewardship of the Earth in the Old Testament." In *Tending the Garden: Essays on the Gospel and the Earth*, edited by Wesley Granberg-Michaelson, 50–65. Grand Rapids: Eerdmans, 1987.

Dyson, Freeman. *Disturbing the Universe.* New York: Harper & Row, 1979.

Earle, Sylvia A. *The World Is Blue: How Our Fate and the Ocean's Are One.* Washington, DC: National Geographic Society, 2009.

"Ecology—Dying to Live Again." *Christianity Today* 16/13 (March 31, 1972) 22–23.

Ellis, George F. R. "Kenosis as a Unifying Theme for Life and Cosmology." In *The Work of Love: Creation as Kenosis*, edited by John Polkinghorne, 107–26. Grand Rapids: Eerdmans, 2001.

Endres, John C. "Praising God the Creator: Praying with Psalms during the Spiritual Exercises of Ignatius." *Ex Auditu* 18 (2002) 93–115.

Engel, J. Ronald, and R. J. S. Montagnes. "Environment." In *Dictionary of Ethics and Theology*, edited by Paul Barry Clarke and Andrew Linzey, 289–94. London: Routledge, 1996.

Erickson, Millard. *Christian Theology.* 2nd ed. Grand Rapids: Baker, 1998.

Estes, J. A., et al. "Complex Trophic Interactions in Kelp Forest Ecosystems." *Bulletin of Marine Science* 74 (2004) 621–38.

Farrington, B. *The Philosophy of Francis Bacon: An Essay on This Development from 1603 to 1609 with New Translations of Fundamental Texts.* Liverpool: Liverpool University Press, 1964.

Fee, Gordon D., and Douglas Stuart. *How to Read the Bible for All Its Worth: A Guide to Understanding the Bible.* 2nd ed. Grand Rapids: Zondervan, 1993.

Fretheim, Terrance E. *God and the World in the Old Testament: A Relational Theology of Creation.* Nashville: Abingdon, 2005.

———. "Nature's Praise of God in the Psalms." *Ex Auditu* 3 (1987) 16–30.

———. "The Reclamation of Creation: Redemption and Law in Exodus." *Interpretation* 45/4 (1991) 354–65.

Fuster, Joaquin M. *Cortex and Mind: Unifying Cognition.* Oxford: Oxford University Press, 2003.

Garcia, Serge M., and Andrew A. Rosenberg. "Food Security and Marine Capture Fisheries: Characteristics, Trends, Drivers, and Future Perspectives." *Philosophical Transactions of the Royal Society B: Biological Sciences* 365/1554 (2010) 2869–80.

Gilkey, Langdon. *Nature, Reality, and the Sacred: The Nexus of Science and Religion.* Minneapolis: Fortress, 1993.

Glacken, Clarence J. *Traces on the Rhodian Shore: Nature and Culture in Western Thought from Ancient Times to the End of the Eighteenth Century*. Berkeley: University of California Press, 1967.
Global Footprint Network. Online: http://www.footprintnetwork.org/.
Goldingay, John. *Old Testament Theology*. Vol. 1, *Israel's Gospel*. Downers Grove, IL: InterVarsity, 2003.
———. *Old Testament Theology*. Vol. 2, *Israel's Faith*. Downers Grove, IL: InterVarsity, 2006.
Goodall, Jane. *The Chimpanzees of Gombe: Patterns of Behavior*. Cambridge: Harvard University Press, 1986.
Gottlieb, Robert. *Reinventing Los Angeles: Nature and Community in the Global City*. Cambridge: MIT Press, 2007.
Gove, Philip Babcock, editor. *Webster's Third New International Dictionary of the English Language Unabridged*. Springfield, MA: Merriam-Webster, 1993.
Green, Joel B. "'Bodies—That Is, Human Lives': A Re-Examination of Human Nature in the Bible." In *Whatever Happened to the Soul? Scientific and Theological Portraits of Human Nature*, edited by Warren S. Brown et al., 149–73. Minneapolis: Fortress, 1998.
———. *Body, Soul, and Human Life: The Nature of Humanity in the Bible*. Grand Rapids: Baker, 2008.
Gregarios, Paulos Mar. "New Testament Foundations for Understanding the Creation." In *Tending the Garden: Essays on the Gospel and the Earth*, edited by Wesley Granberg-Michaelson, 83–92. Grand Rapids: Eerdmans, 1987.
Grenz, Stanley J. *The Social God and the Relational Self: A Trinitarian Theology of the Imago Dei*. Louisville: Westminster John Knox, 2001.
———. *Theology for the Community of God*. Grand Rapids: Eerdmans, 1994.
———, et al. *Pocket Dictionary of Theological Terms*. Downers Grove, IL: InterVarsity, 1999.
Gunton, Colin E. *Christ and Creation: The Didsbury Lectures 1990*. Grand Rapids: Eerdmans, 1992.
———. *The One, the Three and the Many: God, Creation and the Culture of Modernity*. Cambridge: Cambridge University Press, 1993.
———. *The Triune Creator: A Historical and Systematic Study*. Grand Rapids: Eerdmans, 1998.
Gushee, David. "Environmental Ethics: Bringing Creation Care Down to Earth." In *Keeping God's Earth: The Global Environment in Biblical Perspective*, edited by Noah J. Toly and Daniel I. Block, 245–65. Downers Grove, IL: InterVarsity, 2010.
———. "Old-Fashioned Creation Care." *Christianity Today* 51/7 (2007) 51.
Gustafson, James M. *Ethics from a Theocentric Perspective*. Vol. 1, *Theology and Ethics*. Chicago: University of Chicago Press, 1981.
Hall, Douglas John. *Imaging God: Dominion as Stewardship*. Grand Rapids: Eerdmans, 1986.
———. *Professing the Faith: Christian Theology in a North American Context*. Minneapolis: Fortress, 1993.
Hamlin, Christopher, and David M. Lodge. "Beyond Lynn White: Religion, the Contexts of Ecology, and the Flux of Nature." In *Religion and the New Ecology: Environmental Responsibility in a World in Flux*, edited by David M. Lodge and Christopher Hamlin, 1–25. Notre Dame: University of Notre Dame Press, 2006.

Bibliography

Harrington, Jonathan. "Evangelicalism, Environmental Activism, and Climate Change in the United States." *Journal of Religion and Society* 11 (2009). Online: http://moses.creighton.edu/JRS/2009/2009-21.pdf.

Harris, Peter. "Environmental Concern Calls for Repentance and Holiness." In *God's Stewards: The Role of Christians in Creation Care*, edited by Don Brandt, 7–16. Monrovia, CA: World Vision, 2002.

Hart, Ian. "Genesis 1:1—2:3 as a Prologue to the Book of Genesis." *Tyndale Bulletin* 46/2 (1995) 315–36.

Hauerwas, Stanley. *Vision and Virtue: Essays in Christian Ethical Reflection*. Notre Dame: Fides, 1974.

Hixon, M. A. "Carrying Capacity." In *Encyclopedia of Ecology*, edited by Sven Erik Jørgensen and Brian D. Fath, 1:528–30. Amsterdam: Elsevier, 2008.

Henry, Carl F. H. *God, Revelation, and Authority*. Vol. 2, *God Who Speaks and Shows: Fifteen Theses, Part One*. Waco, TX: Word, 1976.

Hoezee, Scott. *Remember Creation: God's World of Wonder and Delight*. Grand Rapids: Eerdmans, 1998.

Høgh-Olesen, Henrik. "Homo Sapiens–Homo Socious: A Comparative Analysis of Human Mind and Kind." In *Human Morality and Sociality: Evolutionary and Comparative Perspectives*, edited by Henrik Høgh-Olesen, 235–71. New York: Macmillan, 2010.

Holling, C. S., and G. K. Meffe. "Command and Control and the Pathology of Natural Resource Management." In *Readings in Ecology*, edited by Stanely I. Dodson et al., 177–90. New York: Oxford University Press, 1999.

Hooper, D. U., et al. "Effects of Biodiversity on Ecosystem Functioning: A Consensus of Current Knowledge." *Ecological Monographs* 75/1 (2005) 3–35.

Horrell, David G. *The Bible and the Environment: Towards a Critical Ecological Biblical Theology*. London: Equinox, 2010.

———. "Introduction." In *Ecological Hermeneutics: Biblical, Historical, and Theological Perspectives*, edited by David G. Horrell et al., 1–12. London: T. & T. Clark, 2010.

Houghton, John. *Global Warming: The Complete Briefing*. 4th ed. Cambridge: Cambridge University Press, 2009.

Huesemann, Michael H. "Can Pollution Problems Be Efffectively Solved by Environmental Science and Technology? An Analysis of Critical Limitations." *Ecological Economics* 37 (2001) 271–87.

Hughes, J. Donald. *An Environmental History of the World: Humankind's Changing Role in the Community of Life*. 2nd ed. London: Routledge, 2009.

Javna, John, Sophie Javna, and Jesse Javna. *50 Simple Things You Can Do to Save the Earth*. New York: Hyperion, 2008.

Jeeves, Malcolm. "Changing Portraits of Human Nature." *Science and Christian Belief* 14 (2002) 3–32.

———. "How Free Is Free? Reflections on the Neuropsychology of Thought and Action." *Science and Christian Belief* 16/2 (2003) 101–22.

———. "The Nature of Persons and the Emergence of Kenotic Behavior." In *The Work of Love: Creation as Kenosis*, edited by John Polkinghorne, 66–89. Grand Rapids: Eerdmans, 2001.

———, and Warren S. Brown. *Neuroscience, Psychology, and Religion: Illusions, Delusions, and Realities about Human Nature*. West Conshohocken, PA: Templeton Foundation, 2009.

Jegen, Mary Evelyn. "The Church's Role in Healing the Earth." In *Tending the Garden: Essays on the Gospel and the Earth*, edited by Wesley Granberg-Michaelson, 93–113. Grand Rapids: Eerdmans, 1987.

Jeremias, Joachim. "Flesh and Blood Cannot Inherit the Kingdom of God (1 Cor. XV.50)." *New Testament Studies* 2 (1956) 151–59.

John Paul II. "Peace with God the Creator, Peace with All of Creation." January 1, 1990. Online: http://www.vatican.va/holy_father/john_paul_ii/messages/peace/documents/hf_jp-ii_mes_19891208_xxiii-world-day-for-peace_en.html.

Jung, Shannon. *We Are Home: A Spirituality of the Environment*. Mahwah, NJ: Paulist, 1993.

Keller, David R., and Frank B. Golley. "Community, Niche, Diversity, and Stability." In *The Philosophy of Ecology: From Science to Synthesis*, edited by David R. Keller and Frank B. Golley, 101–10. Athens: University of Georgia Press, 2000.

———. "Introduction: Ecology as a Science of Synthesis." In *The Philosophy of Ecology: From Science to Synthesis*, edited by David R. Keller and Frank B. Golley, 1–19. Athens: University of Georgia Press, 2000.

Kelsey, David H. *Eccentric Existence: A Theological Anthropology*. Louisville: Westminster John Knox, 2009.

Kostigen, Thomas M. *You Are Here: Exposing the Vital Link between What We Do and What That Does to the Planet*. New York: HarperCollins, 2008.

Kraft, Charles H. "Culture, Worldview and Contextualization." In *Perspectives on the World Christian Movement: A Reader*, edited by Ralph D. Winter and Steven C. Hawthorne, 384–91. 3rd ed. Pasadena, CA: William Carey Library, 1999.

Krebs, Charles J. *Ecology: The Experimental Analysis of Distribution and Abundance*. San Francisco: Benjamin Cummings, 2001.

———, et al. "What Drives the 10-Year Cycle of Snowshoe Hares?" *Bioscience* 51/1 (2001) 25–35.

Langston, Nancy E. "People and Nature: Understanding the Changing Interactions between People and Ecological Systems." In *Ecology*, edited by Stanley I. Dodson et al., 25–76. Oxford: Oxford University Press, 1998.

Laszlo, Ervin. *The Systems View of the World: A Holistic View for Our Time*. Cresskill, NJ: Hampton, 1996.

"Let the Sea Resound." *Christianity Today* 54/8 (August 2010). Online: http://www.christianitytoday.com/ct/2010/august/7.45.html.

Lewis, C. S. *The Problem of Pain*. New York: Macmillan, 1962.

———. *Reflections on the Psalms*. New York: Harcourt Brace Jovanovich, 1958.

Likens, Gene E., and Jerry F. Franklin. "Ecosystem Thinking in the Northern Forest—and Beyond." *Bioscience* 59/6 (2009) 511–13.

Limburg, James. *Psalms*. Louisville: Westminster John Knox, 2000.

Linzey, Andrew. *Animal Theology*. London: SCM, 1995.

———. "Ecological Theology." In *Dictionary of Ethics and Theology*, edited by Paul Barry Clarke and Andrew Linzey, 262–66. London: Routledge, 1996.

Locke, John. *Two Treatises on Government*. Edited by Thomas I. Cook. New York: Hafner, 1947.

Longman, Tremper, III. *How to Read Genesis*. Downers Grove, IL: InterVarsity, 2005.

Lowe, Ben. *Green Revolution: Coming Together to Care for Creation*. Downers Grove, IL: InterVarsity, 2009.

Bibliography

MacIntyre, Alasdair. *Dependent Rational Animals: Why Human Beings Need the Virtues.* Chicago: Carus, 1999.

Malmqvist, Björn, and Simon Rundle. "Threats to the Running Water Ecosystems of the World." *Environmental Conservation* 29/2 (2002) 134–53.

Marten, Gerald G. *Human Ecology: Basic Concepts for Sustainable Development.* London: Earthscan, 2001.

Martin, Raymond, and John Barresi. *The Rise and Fall of Soul and Self: An Intellectual History of Personal Identity.* New York: Columbia University Press, 2006.

McClendon, James William, Jr. *Systematic Theology.* Vol. 1, *Ethics.* Nashville: Abingdon, 1986.

———. *Systematic Theology.* Vol. 3, *Witness.* Nashville: Abingdon, 2000.

McFague, Sallie. *Life Abundant: Rethinking Theology and Economy for a Planet in Peril.* Minneapolis: Fortress, 2001.

———. *A New Climate for Theology: God, the World, and Global Warming.* Minneapolis: Fortress, 2008.

McGinn, Anne Platt. "Combating Malaria." In *Environmental Ethics: Readings in Theory and Application,* edited by Louis P. Pojman and Paul Pojman, 536–62. 5th ed. Belmont, CA: Thomson Wadsworth, 2008.

McGrath, Alister. *The Re-Enchantment of Nature: Science, Religion and the Human Sense of Wonder.* London: Hodder & Stoughton, 2002.

McIntosh, Robert P. *The Background of Ecology: Concept and Theory.* Cambridge: Cambridge University Press, 1985.

McKibben, Bill. "Spitting in the Face of God." *Living Pulpit* 2/2 (1993) 24–25.

McNeill, J. R. *Something New Under the Sun: An Environmental History of the Twentieth-Century World.* New York: Norton, 2000.

Meadows, Donella, Jorgen Randers, and Dennis Meadows. *Limits to Growth: The 30-Year Update.* White River Junction, VT: Chelsea Green, 2004.

Merchant, Carolyn. *The Death of Nature: Women, Ecology, and the Scientific Revolution.* San Francisco: Harper & Row, 1980.

Metzger, James A. *Consumption and Wealth in Luke's Travel Narrative.* Boston: Brill, 2007.

Meye, Robert P. "Invitation to Wonder: Toward a Theology of Nature." In *Tending the Garden: Essays on the Gospel and the Earth,* edited by Wesley Granberg-Michaelson, 30–49. Grand Rapids: Eerdmans, 1987.

Middleton, J. Richard. *The Liberating Image: The Imago Dei in Genesis 1.* Grand Rapids: Brazos, 2005.

Midgley, Mary. *Beast and Man: The Roots of Human Nature.* Rev. ed. London: Routledge, 1995.

———. *The Ethical Primate: Humans, Freedom, and Morality.* London: Routledge, 1994.

Molles, Manuel C., Jr. *Ecology: Concepts and Applications.* 4th ed. New York: McGraw-Hill, 2008.

Moltmann, Jürgen. *God in Creation: A New Theology of Creation and the Spirit of God.* Translated by Margaret Kohl. Minneapolis: Fortress, 1993.

———. "God's Kenosis in the Creation and Consummation of the World." In *The Work of Love: Creation as Kenosis,* edited by John Polkinghorne, 137–51. Grand Rapids: Eerdmans, 2001.

Moo, Douglas J. "Nature in the New Creation: New Testament Eschatology and the Environment." *Journal of the Evangelical Theological Society* 49/3 (2006) 449–88.

Moran, Emilio F. *People and Nature: An Introduction to Human Ecological Relations.* Oxford: Blackwell, 2006.
Moreland, J. P., and Scott B. Rae. *Body and Soul: Human Nature and the Crisis in Ethics.* Downers Grove, IL: InterVarsity, 2000.
Muir, John. *The Wilderness World of John Muir.* Edited by Edwin Way Teale. Boston: Houghton Mifflin, 1954.
Murphy, Charles M. *At Home on Earth: Foundations of a Catholic Ethic of the Environment.* New York: Crossroad, 1989.
Murphy, Nancey. *Bodies and Souls, or Spirited Bodies?* Cambridge: Cambridge University Press, 2006.
———. "Science and Society." In *Systematic Theology* by James William McClendon Jr., 3:99–131. Nashville: Abingdon, 2000.
———, and Warren S. Brown. *Did My Neurons Make Me Do It? Philosophical and Neurobiological Perspectives on Moral Responsibility and Free Will.* Oxford: Oxford University Press, 2007.
———, and George F. R. Ellis. *On the Moral Nature of the Universe: Theology, Cosmology, and Ethics.* Minneapolis: Fortress, 1996.
Myers, Ransom A., and Boris Worm. "Rapid Worldwide Depletion of Predatory Fish Communities." *Nature* 423/6937 (May 15, 2003) 280–83.
———, et al. "Why Do Fish Stocks Collapse? The Example of Cod in Atlantic Canada." *Ecological Applications* 7/1 (1997) 91–106.
Nash, James A. "Christianity's Ecological Reformation." In *The Encyclopedia of Religion and Nature*, edited by Bron R. Taylor, 372–75. London: Thoemmes Continuum, 2005.
———. *Loving Nature: Ecological Integrity and Christian Responsibility.* Nashville: Abingdon, 1991.
———. "Toward the Ecological Reformation of Christianity." *Interpretation* 50/1 (1996) 5–15.
Nash, Roderick Frazier. *The Rights of Nature: A History of Environmental Ethics.* Madison: University of Wisconsin Press, 1989.
———. *Wilderness and the American Mind.* 4th ed. New Haven: Yale University Press, 2001.
National Environmental Education Foundation. *National Environmental Report Card.* 1999. Online: http://www.neefusa.org/pdf/roper/99reportcard.pdf.
National Oceanic and Atmospheric Administration. "Mauna Loa Annual Mean Data." April 5, 2012. Online: ftp://ftp.cmdl.noaa.gov/ccg/co2/trends/co2_annmean_mlo.txt.
Naylor, Rosamond L., et al. "Effect of Aquaculture on World Fish Supplies." *Nature* 405 (June 29, 2000) 1017–24.
Neff, David. "Second Coming Ecology." *Christianity Today* 52/7 (July 2008). Online: http://www.christianitytoday.com/ct/2008/july/23.35.html.
Newsom, Carol A. "The Moral Sense of Nature: Ethics in the Light of God's Speech to Job." *The Princeton Seminary Bulletin* 15/1 (1994) 9–27.
Niebuhr, Reinhold. *The Nature and Destiny of Man: A Christian Interpretation.* New York: Scribner's, 1948.
Northcott, Michael S. *The Environment and Christian Ethics.* Cambridge: Cambridge University Press, 1996.

Bibliography

———. "Sustaining Ethical Life in the Anthropocene." In *Creation in Crisis: Christian Perspectives on Sustainability*, edited by Robert S. White, 225–40. London: SPCK, 2009.
Nowak, Martin A., and Karl Sigmund. "How Populations Cohere: Five Rules for Cooperation." In *Theoretical Ecology: Principles and Applications*, edited by Robert M. May and Angela R. McLean, 7–34. 3rd ed. Oxford: Oxford University Press, 2007.
O'Brien, Kevin J. *An Ethics of Biodiversity: Christianity, Ecology, and the Variety of Life*. Washington, DC: Georgetown University Press, 2010.
O'Donovan, Oliver. *Resurrection and the Moral Order: An Outline of Evangelical Ethics*. 2nd ed. Grand Rapids: Eerdmans, 1994.
Odum, Eugene P. "The Emergence of Ecology as a New Integrative Discipline." In *The Philosophy of Ecology: From Science to Synthesis*, edited by David P. Keller and Frank B. Golley, 194–203. Athens: University of Georgia Press, 2000.
———. "The New Ecology." *Bioscience* 14/7 (1964) 14–16.
———, and Gary W. Barrett. *Fundamentals of Ecology*. 5th ed. Belmont, CA: Brooks/Cole, 2005.
Oelschlaeger, Max. *Caring for Creation: An Ecumenical Approach to the Environmental Crisis*. London: Yale University Press, 1994.
O'Neill, R. V. "Perspectives on Economics and Ecology." *Ecological Applications* 6/4 (1996) 1031–33.
Orr, David W. "Retrospect and Prospect: The Unbearable Lightness of Conservation." *Conservation Biology* 23/6 (2009) 1349–51.
Palmer, Clare. "An Overview of Environmental Ethics." In *Environmental Ethics: An Anthology*, edited by Andrew Light and Holmes Rolston III, 15–37. Oxford: Blackwell, 2003.
———. "Stewardship: A Case Study in Environmental Ethics." In *Environmental Stewardship: Critical Perspectives—Past and Present*, edited by R. J. Berry, 63–75. London: T. & T. Clark, 2006.
Pannenberg, Wolfhart. *Anthropology in Theological Perspective*. Translated by Matthew J. O'Connell. Philadelphia: Westminster, 1985.
———. *Systematic Theology*. Vol. 2. Translated by Geoffrey W. Bromiley. Grand Rapids: Eerdmans, 1994.
Passingham, Richard. *What Is Special about the Human Brain?* Oxford: Oxford University Press, 2008.
Patz, Jonathan A., and R. Sari Kovats. "Hotspots in Climate Change and Human Health." *British Medical Journal* 324/7372 (November 9, 2002) 1094–98. Online: http://www.ncbi.nlm.nih.gov/pmc/articles/PMC1124582/.
Pauly, Daniel, et al. "Fishing Down Marine Food Webs." *Science* 279/5352 (February 6, 1998) 860–63.
Peacocke, Arthur. "The Cost of New Life." In *The Work of Love: Creation as Kenosis*, edited by John Polkinghorne, 21–42. Grand Rapids: Eerdmans, 2001.
———. *Creation and the World of Science: The Re-Shaping of Belief*. Oxford: Oxford University Press, 1979.
Peterson, Anna L. *Being Human: Ethics, Environment, and Our Place in the World*. Berkeley: University of California Press, 2001.
———. "In and of the World? Christian Theological Anthropology and Environmental Ethics." *Journal of Agricultural and Environmental Ethics* 12/3 (2000) 237–61.

Phillips, J. B. *Your God Is Too Small*. New York: Simon & Schuster, 1997.
Pickett, Steward T. A., et al. "The New Paradigm in Ecology: Implications for Conservation Biology above the Species Level." In *Conservation Biology: The Theory and Practice of Nature Conservation, Preservation and Management*, edited by Peggy L. Fiedler and Subodh K. Jain, 65–88. New York: Routledge Chapman & Hall, 1992.
Polkinghorne, John. "The Friendship of Science and Religion." Lecture given at Point Loma Nazarene University, San Diego, CA, November 14, 2010.
———. "Kenotic Creation and Divine Action." In *The Work of Love: Creation as Kenosis*, edited by John Polkinghorne, 90–106. Grand Rapids: Eerdmans, 2001.
———. *Scientists as Theologians: A Comparison of the Writings of Ian Barbour, Arthur Peacocke, and John Polkinghorne*. London: SPCK, 1997.
Ponting, Clive. *A New Green History of the World: The Environment and the Collapse of Great Civilizations*. Rev. ed. New York: Penguin, 2007.
Powers, Mary E., et al. "Challenges in the Quest for Keystones." *Bioscience* 46/8 (1996) 609–20.
Rad, Gerhard von. *Genesis: A Commentary*. Rev. ed. Philadelphia: Westminster, 1972.
———. "*eikon.*" In *Theological Dictionary of the Old Testament*, edited by Gerhard Kittel, translated by Geoffrey W. Bromiley, 3:390. Grand Rapids: Eerdmans, 1964.
———. *Old Testament Theology*. Vol. 1, *The Theology of Israel's Historical Traditions*. Edinburgh: Oliver & Boyd, 1965.
Rasmussen, Larry. "Creation, Church, and Christian Responsibility." In *Tending the Garden: Essays on the Gospel and the Earth*, edited by Wesley Granberg-Michaelson, 114–31. Grand Rapids: Eerdmans, 1987.
———. *Earth Community, Earth Ethics*. Maryknoll, NY: Orbis, 1996.
———. "Ecology and Morality: The Challenge to and from Christian Ethics." In *Religion and the New Ecology: Environmental Responsibility in a World in Flux*, edited by David M. Lodge and Christopher Hamlin, 246–78. Notre Dame: University of Notre Dame Press, 2006.
Rogers, A. D., and D. d'A. Laffoley. *International Earth System Expert Workshop on Ocean Stresses and Impacts: Summary Report*. International Program on the State of the Oceans, Oxford, 2011. Online: http://www.stateoftheocean.org/pdfs/1906_IPSO-LONG.pdf.
Rojstaczer, Stuart, Shannon M. Sterling, and Nathan J. Moore. "Human Appropriation of Photosynthesis Products." *Science* 294/5551 (December 21, 2001) 2549–52.
Raven, Peter H. "Foreword." In *Religion and the New Ecology: Environmental Responsibility in a World in Flux*, edited by David M. Lodge and Christopher Hamlin, vii–x. Notre Dame: University of Notre Dame Press, 2006.
Rolston, Holmes, III. "Does Nature Need to Be Redeemed?" *Zygon* 29/2 (1994) 205–29.
———. "Kenosis and Nature." In *The Work of Love: Creation as Kenosis*, edited by John Polkinghorne, 43–65. Grand Rapids: Eerdmans, 2001.
———. *Philosophy Gone Wild: Environmental Ethics*. Buffalo, NY: Prometheus, 1986.
Rossi, Vincent. "Theocentrism: The Cornerstone of Christian Ecology." *Epiphany* 6/1 (1985) 8–14.
Sabin, Scott C. *Tending to Eden: Environmental Stewardship for God's People*. Valley Forge, PA: Judson, 2010.
Sagoff, Mark. *The Economy of the Earth: Philosophy, Law, and the Environment*. 2nd ed. Cambridge: Cambridge University Press, 2008.

Bibliography

Santmire, H. Paul. *Brother Earth: Nature, God and Ecology in Time of Crisis*. New York: Thomas Nelson, 1970.

———. "From Consumerism to Stewardship: The Troublesome Ambiguities of an Attractive Option." *Dialog* 49/4 (2010) 332–39.

———. "Historical Dimensions of the American Crisis." In *Western Man and Environmental Ethics: Attitudes toward Nature and Technology*, edited by Ian G. Barbour, 66–92. Reading, MA: Addison-Wesley, 1973.

———. "Partnership with Nature according to the Scriptures: Beyond the Theology of Stewardship." *Christian Scholars Review* 32/4 (2003) 381–412.

———. *The Travail of Nature: The Ambiguous Ecological Promise of Christian Theology*. Philadelphia: Fortress, 1985.

Schaeffer, Francis A. *Pollution and the Death of Man: The Christian View of Ecology*. Wheaton, IL: Tyndale, 1970.

Schneider, E. D., and J. J. Kay. "Life as a Manifestation of the Second Law of Thermodynamics." In *Readings in Ecology*, edited by Stanley I. Dodson, et al., 435–61. Oxford: Oxford University Press, 1999.

Schwöbel, Christoph. "God, Creation and the Christian Community: The Dogmatic Basis of a Christian Ethic of Createdness." In *The Doctrine of Creation: Essays in Dogmatics, History and Philosophy*, edited by Colin E. Gunton, 149–75. Edinburgh: T. & T. Clark, 1997.

Scorgie, Glen G. *A Little Guide to Christian Spirituality: Three Dimensions of Life with God*. Grand Rapids: Zondervan, 2007.

———. "Wonder and the Revitalization of Evangelical Theology." *Crux* 26/4 (1990) 19–25.

Sears, Paul B. "Ecology—A Subversive Subject." *Bioscience* 14/7 (1964) 11–13.

Shults, F. LeRon. *Reforming Theological Anthropology: After the Philosophical Turn to Relationality*. Grand Rapids: Eerdmans, 2003.

Simmons, I. G. *Global Environmental History*. Chicago: University of Chicago Press, 2008.

Simon, Julian L. *The Ultimate Resource 2*. Princeton: Princeton University Press, 1996.

Sittler, Joseph. *Evocations of Grace: The Writings of Joseph Sittler on Ecology, Theology, and Ethics*. Edited by Steven Bouma-Prediger and Peter Bakken. Grand Rapids: Eerdmans, 2000.

———. "A Theology of the Earth." *The Christian Scholar* 37/3 (1954) 367–74.

Sklair, Leslie. *Sociology of the Global System*. New York: Harvester/Wheatsheaf, 1991.

Sleeth, Emma. *It's Not Easy Being Green: One Student's Guide to Serving God and Saving the Planet*. El Cajon, CA: Youth Specialties, 2008.

Sleeth, Matthew. *Hope for Creation Guidebook*. Grand Rapids: Zondervan, 2010.

Sleeth, Nancy. *Go Green, $ave Green: A Simple Guide to Saving Time, Money, and God's Green Earth*. Carol Stream, IL: Tyndale, 2009.

Smith, Christian. *Moral, Believing Animals: Human Personhood and Culture*. Oxford: Oxford University Press, 2003.

Smith, Stephen M. "Theology of Hope." In *Evangelical Dictionary of Theology*, edited by Walter A. Elwell, 533–34. Grand Rapids: Baker, 1984.

Snodgrass, Klyne. "Jesus and Money—No Place to Hide and No Easy Answers." *Word & World* 30/2 (2010) 135–43.

Solomon, Steven. *Water: The Epic Struggle for Wealth, Power, and Civilization*. New York: HarperCollins, 2010.

Sorrell, Roger D. *St. Francis of Assisi and Nature: Tradition and Innovation in Western Christian Attitudes toward the Environment*. Oxford: Oxford University Press, 1988.

Soulé, Michael. E. "What Is Conservation Biology?" *Bioscience* 35/11 (1985) 727–34.

Southgate, Christopher. *The Groaning of Creation: God, Evolution, and the Problem of Evil*. Louisville: Westminster John Knox, 2008.

Speth, James Gustave. *Red Sky at Morning: America and the Crisis of the Global Environment*. New Haven: Yale University Press, 2004.

Staats, Reinhart. "Asceticism." In *The Encyclopedia of Christianity*, edited by Erwin Fahlbusch et al., translated by Geoffrey W. Bromiley, 1:131–34. Grand Rapids: Eerdmans, 1999.

Stassen, Glen H., and David P. Gushee. *Kingdom Ethics: Following Jesus in Contemporary Context*. Downers Grove, IL: InterVarsity, 2003.

Stern, Nicholas, et al. *Stern Review on the Economics of Climate Change*. HM Treasury (UK). 2006. Online: http://webarchive.nationalarchives.gov.uk/+/http://www.hm-treasury.gov.uk/stern_review_report.htm.

Stone, Lawson G. "The Soul: Possession, Part, or Person? The Genesis of Human Nature in Genesis 2:7." In *What about the Soul? Neuroscience and Christian Anthropology*, edited by Joel B. Green, 47–61. Nashville: Abingdon, 2004.

Stott, John. *The Birds Our Teachers: Lessons From a Lifelong Bird Watcher*. Grand Rapids: Baker, 2008.

Strier, Karen B. "Primate Behavioral Ecology: From Ethnography to Ethology and Back." *American Anthropologist* 105/1 (2003) 16–27.

Stuhlmacher, Peter. "The Ecological Crisis as a Challenge for Biblical Theology." *Ex Auditu* 3 (1987) 1–15.

Sukhdev, Pavan, et al. *TEEB (2010): The Economics of Ecosystems and Biodiversity: Mainstreaming the Economics of Nature: A Synthesis of the Approach, Conclusions, and Recommendations of TEEB*. N.p.: TEEB, 2010. Online: http://www.teebweb.org/LinkClick.aspx?fileticket=bYhDohL_TuM%3D.

Tarlock, A. Dan. "The Nonequilibrium Paradigm in Ecology and the Partial Unraveling of Environmental Law." *Loyola of Los Angeles Law Review* 27/3 (1994) 1121–44.

Taylor, Charles. *Sources of the Self: The Making of the Modern Identity*. Cambridge: Harvard University Press, 1989.

Taylor, Daniel. *The Myth of Certainty: The Reflective Christian and the Risk of Commitment*. Waco, TX: Jarrell, 1986.

Taylor, Paul W. "The Ethics of Respect for Nature." In *Environmental Ethics: An Anthology*, edited by Andrew Light and Holmes Rolston III, 74–84. Oxford: Blackwell, 2003.

Teale, Edwin Way. *The Strange Lives of Familiar Insects*. New York: Dodd Mead, 1962.

Thompson, Geoff. "'Remaining Loyal to the Earth': Humanity, God's Other Creatures, and the Bible in Karl Barth." In *Ecological Hermeneutics: Biblical, Historical, and Theological Perspectives*, edited by David G. Horrell et al., 181–95. London: T. & T. Clark, 2010.

Thompson, John N., et al. "Frontiers in Ecology." *Bioscience* 51/1 (2001) 15–24.

Tilman, G. David. "Global Environmental Impacts of Agricultural Expansion: The Need for Sustainable and Efficient Practices." *Proceedings of the National Academy of Science* 96 (May 25, 1999) 5995–6000.

Bibliography

Toulmin, Stephen. "Cosmology as Science and as Religion." In *On Nature*, edited by Leroy S. Roumer, 27–41. Notre Dame: University of Notre Dame Press, 1984.

———. "Nature and Nature's God." *The Journal of Religious Ethics* 13/1 (1985) 37–52.

———. "Religion and the Idea of Nature." In *Religion, Science, and Public Policy*, edited by Frank T. Birtel, 67–78. New York: Crossroad, 1987.

Towner, W. Sibley. "The Future of Nature." *Interpretation* 50/1 (1996) 27–35.

Turner, Monica G. "Landscape Ecology." In *Ecology*, by Stanley I. Dodson et al., 78–121. Oxford: Oxford University Press, 1998.

Valerio, Ruth. *"L" Is for Lifestyle: Christian Living that Doesn't Cost the Earth*. Rev. ed. Nottingham, UK: InterVarsity, 2008.

Van Dyke, Fred. *Between Heaven and Earth: Christian Perspectives on Environmental Protection*. Santa Barbara, CA: Praeger, 2010.

———. "Bridging the Gap: Christian Environmental Stewardship and Public Environmental Policy." *Trinity Journal* 18/2 (1997) 139–72.

———. *Conservation Biology: Foundations, Concepts, Applications*. 2nd ed. New York: Springer, 2010.

———. "Planetary Economies and Ecologies: The Christian World View in Recent Literature." *Perspectives on Science and the Christian Faith* 40 (1988) 66–71. Online: http://www.asa3.org/ASA/PSCF/1988/PSCF6-88VanDyke.html.

———, et al. *Redeeming Creation: The Biblical Basis of Environmental Stewardship*. Downers Grove, IL: InterVarsity, 1996.

Van Houtan, Kyle S., and Stuart L. Pimm. "The Various Christian Ethics of Species Conservation." In *Religion and the New Ecology: Environmental Responsibility in a World in Flux*, edited by David M. Lodge and Christopher Hamlin, 116–47. Notre Dame: University of Notre Dame Press, 2006.

Van Huyssteen, J. Wentzel. *Alone in the World? Human Uniqueness in Science and Theology*. Grand Rapids: Eerdmans, 2006.

Vitousek, Peter M. "Beyond Global Warming: Ecology and Global Change." *Ecology* 75/7 (1994) 1861–76.

———, and Harold A. Mooney. "Human Domination of Earth's Ecosystems." *Science* 277/5325 (July 25, 2007) 494–99.

———, et al. "Human Alteration of the Global Nitrogen Cycle." *Ecological Applications* 7/3 (1997) 737–50.

Wackernagel, M., and J. Kitzes. "Ecological Footprint." In *Encyclopedia of Ecology*, edited by Sven Erik Jørgensen and Brian D. Fath, 1:1031–37. Amsterdam: Elsevier, 2008.

———, and William Rees. *Our Ecological Footprint: Reducing Human Impact on the Earth*. Gabriola Island, BC: New Society, 1996.

Westermann, Claus. *Elements of Old Testament Theology*. Translated by Douglas W. Scott. Atlanta: John Knox, 1982.

———. *Genesis 1–11: A Commentary*. Translated by John J. Scullion. Minneapolis: Augsburg, 1984.

White, Lynn, Jr. "The Historical Roots of Our Ecological Crisis." *Science* 155/3767 (March 10, 1967) 1203–7.

White, Peter S. "Disturbance, the Flux of Nature, and Environmental Ethics at the Multipatch Scale." In *Religion and the New Ecology: Environmental Responsibility in a World in Flux*, edited by David M. Lodge and Christopher Hamlin, 176–98. Notre Dame: University of Notre Dame Press, 2006.

Whitney, Elspeth. "Christianity and Changing Concepts of Nature: An Historical Perspective." In *Religion and the New Ecology: Environmental Responsibility in a World in Flux*, edited by David M. Lodge and Christopher Hamlin, 26–52. Notre Dame: University of Notre Dame Press, 2006.

Wilkinson, Loren. "Cosmic Christology and the Christian's Role in Creation." *Christian Scholar's Review* 11/1 (1981) 18–40.

———. "Eco-Myths: There Is Nothing Christians Can Do." *Christianity Today* 38/4 (April 4, 1994) 31–33.

———. "The Meaning of 'Value' in Biodiversity." *Crux* 38/2 (2002) 8–18.

———. "New Age, New Consciousness, and the New Creation." In *Tending the Garden: Essays on the Gospel and the Earth*, edited by Wesley Granberg-Michaelson, 6–29. Grand Rapids: Eerdmans, 1987.

———. "The New Story of Creation: A Trinitarian Perspective." *Crux* 30/4 (1994) 26–36.

Williams, Patricia A. *Doing Without Adam and Eve: Sociobiology and Original Sin*. Minneapolis: Fortress, 2001.

Wilson, Edward O. *The Future of Life*. New York: Vintage, 2002.

Wright, Christopher J. H. "'The Earth is the Lord's': Biblical Foundations for Global Ecological Ethics and Mission." In *Keeping God's Earth: The Global Environment in Biblical Perspective*, edited by Noah J. Toly and Daniel I. Block, 216–41. Downers Grove, IL: InterVarsity, 2010.

———. *Old Testament Ethics for the People of God*. Downers Grove, IL: InterVarsity, 2004.

Zeman, Adam. "Does Consciousness Spring from the Brain? Dilemmas of Awareness in Practice and in Theory." In *Frontiers of Consciousness*, edited by Lawrence Weiskrantz and Martin Davies, 289–322. Oxford: Oxford University Press, 2008.

Index

abiotic, 251
Abram, 184–85
Achtemeier, Elizabeth, 31, 96, 101, 131
agriculture, 68
Ahab, 87–88
alienation. *See* separation from nature
Allen, Diogenes, ix
altruism, 61, 72
ancient peoples, 32
Anderson, Bernard W., 90, 98
animals, 57, 60;
    moral-like behavior in, 60–62
anthropocentrism, 251
anthropology, 14, 121–22, 251
Aquinas, Thomas, 36
Aristotle, 33–34, 36, 65
asceticism, 214
Augustine, 34–35, 120
autotroph, 251

Bacon, Francis, 41–42
Barresi, John, 33, 34, 41, 45
Barrett, Gary, 163, 167–69, 186, 192–93, 196, 251, 255–256
Bauckham, Richard, 27, 49, 52, 74, 84, 89, 90, 97, 108–10, 114, 125, 128, 131–32, 136, 142, 145, 185, 207, 213, 220–22, 224–25
Bauer, Walter, 93, 207, 210
Beisner, E. Calvin, 37
Belovsky, Gary E., 155, 187–88, 196
Berkouwer, G. C., 144
Bernstein, Brock, 159
Berry, R. J., 56, 59, 65, 120, 135, 196
Berry, Thomas, 51
Berry, Wendell, 85, 206
biocentrism, 251

biodiversity, 175–76, 251
biological magnification, 172–73
biota, 251
biotic, 251
Block, Daniel I., 96, 107–8
Bodley, John H., 241
body, 2, 14, 100–1
Bonhoeffer, Dietrich, 214
Bookless, Dave, 144, 216, 247
Bouma-Prediger, Steven, 52, 79, 81, 89, 95, 103, 206, 207, 210, 241
brain, 63–64
Bratton, Susan P., 51, 133, 135, 183–84, 190, 200, 209, 211, 238
Brown, Warren, 33, 55, 56, 65, 66, 124, 125
Brueggemann, Walter, 87, 92, 123, 220
Bruno, Giordano, 53

Cahill, Thomas, 249
Callicott, J. Baird, 194
Calvin, John, 37, 222, 231
Campbell, Neil A., 162, 170
capitalism. *See* economics
carbon cycle, 165–66
caretaking mandate, 132
carrying capacity, 180–82
Carson, Rachel, 173
chimpanzees, 58–59, 67
cities, 49–51
Clark, Anna M., 31
classical world, 32–35
Clifford, Anne M., 41, 142
climate change, 10–11, 157, 161, 165, 181
climax community, 252
cognition, 58–59, 66–67

274

Index

Cohen, Joel E., 177, 181–82
command-and-control, 41–42, 79, 86, 161–62, 199–200
communication, 65–66
complexity, 125, 155–56
Conradie, Ernst M., 63, 67, 74, 98, 105, 132–33, 205, 250
contingency, 101–3
Costanza, Robert, 162, 176, 225–26, 240–42, 244–45
creation, 23, 53, 94–95, 252
    contingency of, 101–2
    doctrine of, 84–86
    holistic view of, 113–15
creatureliness, 139
creatures, dignity and value of, 127
culture, 59–60, 68, 77, 252

Daly, Herman E., 239–40
Damasio, Antonio, 55–56
Darwin, Charles, 45–46
David, 219
Davis, Ellen F., 17
DDT, 172–73
demographic transition, 236–37, 252
Descartes, René, 40–41
DeWitt, Calvin, 31, 108, 162
Diamond, Jared, 71, 194, 243
dichotomism. *See* dualism, body-soul
disequilibrium, 252
disturbance, 197–98, 253
diversity. *See* biodiversity.
docetism, 117
dolphins, 58, 59
dominion mandate, 36, 42, 122, 128–38
    as dominion over ourselves, 133, 211–12
    as gift, 131
    as God's blessing, 134–35
    in the image of Christ, 135–38
    moderation of, 133–34
    power of, 130–31
    as praise of God, 224–25
    stewardship and, 131–32;
Dorn, Harold, 178
Douglas, David, 85, 92

dualism, ix–x, 253
    body-soul, 1, 14, 33, 40–41, 252
    spirit-physical, 12–13, 17, 40–41
Dubos, René, 191, 206
Dunn, James, 100–101, 146, 250
dust, 1, 118
Dyrness, William, 143, 208
Dyson, Freeman, xi

Earle, Sylvia A., 11
earth, 14, 18
Echlin, Edward, 6
eco-consciousness, 230–31, 248
ecological economics, 244–46
    education, 231–33
    footprint, 183–85
    knowledge, 156–58
    realities, 5, 24, 158–94, 253
    systems, 155–56, 188
    ecological life, 62–63
Ecological Problem, 7–12, 13, 18, 21, 203, 253
ecology, 27, 253
    as interdisciplinary science, 155
    new, 194–200
    as young science, 154–55
economics, ecology and, 239–44
eco-physical beings, humans as, xiii, 5, 81
ecosphere, 4, 254
ecosystem 254
    stability, 175–76
    services, 176–77
Ekins, Paul, 241
Ellis, George, 102, 217, 220
Elton, Charles, 195
empiricism, 43
Endres, John C., 224
energy flow and exchange, 166–68
energy pyramid, 167
Engel, J. Ronald, 32
"enlightened self-interest," 203–7
Enlightenment, 43–44, 254
entropy, 185–90, 254
environment, environmental, 254

Erickson, Millard, 15, 23–24, 91, 121–23, 139, 141, 253
eschatology, 255
ethical species, humans as the, 73–74, 228
ethics, 255. *See also* morality
ethology, 255
eutrophication, 164, 255
evolution, 45–46

Farley, Joshua, 239–40
farming, 159–60
Fee, Gordon D., 80, 81
feedback loops, 158–62
    time lags in, 161–62
fertility, 255
fire, 71
fisheries, 9
flux, 195–96
food, 160
food webs, 168–75
    grazing, 169
    detritus, 169
footprint. *See* ecological footprint
form, forms, 34
fossil fuels, 49
Francis of Assisi, 57, 87, 149–50
Franklin, Jerry R., 230, 236
free choice, 69
Fretheim, Terrance E., 92, 114, 145, 221, 224
Fuster, Joaquin M., 16, 57, 64
future-planning, 69

Gage, Phineas P., 55–56
Gandhi, Mahatma, 219–20
Garcia, Serge M., 9
gift, 103
Gilkey, Langdon, 48
Glacken, Clarence J., 37, 46, 97, 178
global warming. *See* climate change
Gnosticism, 33
God, 18, 76, 82–95, 102–3, 149–50
    Creator, 83–86
    delight in creation, 92–95
    immanence of, direct relationship with all creatures, 90–92
    of kenosis, 93–95
    King of creation
    Owner of all, 86–89
    Provider, 89–90
    as unique, transcendent being, 82–83
Goldingay, John, 129, 134
Golley, Frank B., 195, 256
Goodall, Jane, 58, 60
Great Chain of Being, 34–36
Green, Joel B., 56, 61, 73, 99–100, 125, 141, 143
green living, 246–47
Gregarios, Paulos Mar, 116–17
Grenz, Stanley, 216, 221, 253, 255
growth, economic, 240–41
Gunton, Colin, 30, 98, 104, 120, 124, 127, 146
Gushee, David, 21, 29, 147, 151, 211, 219
Gustafson, James, 151

habitat, 256
Hall, Douglas John, 6, 96, 98, 118, 140
Harrington, Jonathan, 235
Harris, Peter, 84–85, 115
Hart, Ian, 122, 131
Hauerwas, Stanley, 7, 248
Henry, Carl F. H., 31–32, 82
Hixon, M. A., 183
Hoezee, Scott, 92, 222, 232
Høgh-Olesen, Henrik, 68
holism, 256
Holling, C. S., 192, 199
Hooper, D. U., 175–76
Horrell, David G., 79, 110
Houghton, John, 165, 239
Huesemann, Michael H., 17
Hughes, J. Donald, 32, 35, 189
human constitution, 256. *See also* anthropology
humanism, 19
humans, humanity, 95–109
    as animals, 96

contingency of, 101–3
  embeddedness of, 103–5
  limitations of, 105
  physicality of, 99–101
  relationality of, 105–9
  uniqueness of, 74–76
humility, 207–10

image of God, 118–28
  as basis of human dignity, 126–27
  as basis for human uniqueness, 120–21
  functional interpretation of, 122–23
  as gift, 125–26; gives power, 126
  goal of, 127–28
  holistic interpretation of, 120
  physical nature of, 119
  relational interpretation of, 123–25
  substantive interpretation of, 121–22
immortality of the soul, 14, 33
incarnation, 116
individualism, 44–45, 234
instrumentalism, 41–42
interpretation of Scripture, 76–82

Javna, John, 247
Jeeves, Malcolm, 33, 55, 56, 59, 65, 66, 78, 120, 125, 145
Jegen, Mary, 220
Jeremias, Joachim, 146
Jesus Christ, 18, 22, 88, 93–94, 116–17, 127–28, 135–38, 147, 197, 210, 215–17, 224, 250
Job, 109–10
John, 95
John Paul II, 219
Jung, Shannon, 5, 103
justice, 217–20, 237–38

Kay, J. J., 187–88
Keller, David R., 195, 256
Kelsey, David H., 103, 116, 127–28
kenosis, 92–95, 208, 216–17

Kitzes, J., 8, 183–84, 218
Kostigen, Thomas, 171
Kovats, R. Sari, 218
Kraft, Charles, 252, 258
Krebs, Charles J., 27, 159, 166, 171, 174, 180, 182, 191
Kuhn, Thomas, 36, 46

Lafoley, D. d'A., 9
Lake Victoria, 174–75
land, 87–89, 106–7
landscape, 256
Langston, Nancy E., 258
language. *See* communication
Laszlo, Ervin, 189
Legacy Movement, 247
Lewis, C. S., 62, 85, 115
lifestyle, 233–36
Likens, Gene E., 230, 236
limits, principle of, 191–94, 237, 241–43, 245
Linzey, Andrew, 34, 217
Locke, John, 91
Longman, Tremper, 83–84
Lot, 184–85
love, 72–73, 94–95, 102
Lowe, Ben, 233
lynx, 158–59

MacIntyre, Alasdair, 58, 59, 61, 67, 68, 74, 200
malaria, 203
Malmqvist, Björn, 10
management, 198–99
Martin, Raymond, 33, 34, 41, 45
material. *See* physical
McClendon, James, 18, 96
McFague, Sallie, 23, 67
McGrath, Alister, 32, 35, 47, 53, 140, 214
McKibben, Bill, 110
McNeill, J. R., 7, 239–40, 243
Meadows, Donella, 182
mechanism, philosophical, 39
Meffe, G. K., 192, 199

Merchant, Carolyn, 39
metacognition. *See* cognition
Metzger, James A., 82, 215
Meye, Robert P., 223
Middle Ages, 35–37
Middleton, J. Richard, 118, 123, 126, 136–37
Midgley, Mary, 39, 60
migration, 71
moderation, 133–34
modern period, 38
modern turn to the self, 44
Molles, Manuel, 5, 159, 174, 179, 251–52, 256
Moltmann, Jürgen, 94–97, 120, 145, 208
monism, 15, 256
Montagnes, R. J. S., 32
Moo, Douglas, 78–79, 145, 148, 215, 217
Mooney, Harold A., 8
morality, 69–70, 81, 202, 256
Moran, Emilio F., 32
Moreland, J. P., 103
Muir, John, 31
Murphy, Charles M., 53
Murphy, Nancey C., ix–xii, 2, 34, 56, 57, 102, 124, 220
mutualism, 72
Myers, Ransom A., 9, 193

Naboth, 87–88
Nagel, Thomas, 2
Nash, James A., 58, 64, 115, 144, 194, 212, 245
Nash, Roderick, 45
National Environmental Education Foundation, 232
National Oceanic and Atmospheric Administration, 10
natural evil, 26–27
nature, 256
neuroscience, 16, 55–57, 256
new ecology. *See* ecology, new
Newton, Isaac, 38–39
niche, 256
Niebuhr, Reinhold, 140–41

Nigeria, 218
Nile perch, 174–75
nitrogen cycle, 163–65
nonequilibrium, 195–96
nonlinear, nonlinearity, 256
nonnative species, 9, 174–75
Northcott, Michael, 30
Nowak, Martin A., 61

O'Brien, Kevin, 234
O'Donovan, Oliver, 24–25, 134
Odum, Eugene, 158, 163, 167–69, 186, 192–93, 196, 251, 255–256
Oelschlaeger, Max, 17, 20
oikos, 23
O'Neill, R. V., 238
ontology, ontological, 257
Orr, David W. 27, 31, 157
otters, 173–74
overfishing, 9, 174, 192–93
overweight and obesity, 160–61
Pannenberg, Wolfhart, 16, 25, 97, 133, 137
Pascal, Blaise, 54
Passingham, Richard, 59, 63–67, 69
patch, 257
Patz, Jonathan A., 218
Paul, Apostle, 23, 28, 79, 80, 100, 104, 147–49, 223
Pauly, Daniel, 9
Peacocke, Arthur, 57, 116, 121, 129, 136, 217
Peter, Apostle, 100, 250
Peterson, Anna, xiii, 28, 45, 129
Phillips, J. B., 86
photosynthesis, 11
physical features, 58
physical spirituality. *See* spirituality, physical
physical, physicality, 4, 17, 99–101, 257
physicalism, 15
Pickett, Steward, 195
Pim, Stuart L., 11
place, 51–52
Plato, ix–x, 33
Polanyi, Michael, 46

Polkinghorne, John, 17, 21, 56, 208
population dynamics, 177–85, 257
  human, 181–83, 236–39
  logistical model of, 179–81
post-modern movement, 47
power, 130–31
Powers, Mary, 174
Poyourow, Joanne, 247
praising God with creation, 220–25
pride, 216
primary producer. *See* autotroph
primatology, 257
property, 86–89
providence. *See* God, Provider.
public policy, 235–36
Puritans, 42–43

Quiverfull Movement, 236

Rae, Scott B., 103
Rasmussen, Larry, 5, 44, 141, 233
Raven, Peter H., 248
redemption, holistic, 144–48
reductionism, 257
Reece, Jane B., 162, 170
Rees, William, 172, 179, 182–83
Reformation, 37–38
relationality, 64, 105–6, 108, 123–25
religion, 17, 70–71
Renaissance, 37–38
resistance, 214–15
resource recycling, 162–66
resurrection, x, 14, 249
Rogers, A. D., 9
Rolston, Holmes, 154, 217
romantic movement, 47
Rosenberg, Andrew A., 9
Rundle, Simon, 10

Sabin, Scott, 205
Sagoff, Mark, 242
San Diego, 50, 161, 201
Santmire, H. Paul, 7, 35, 39, 42, 123, 130, 132, 221, 232, 241

Schaeffer, Francis, 7, 21, 111, 130, 147, 210
Schneider, E. D., 187–88
Schwöbel, Christoph, 102
science, 21, 77–78
Scorgie, Glen, 124, 131, 215, 230, 246, 249, 258
Sears, Paul, 244
second law of thermodynamics. *See* entropy.
self-limitation, 94–95, 210–16
separation from nature, 6, 30–53, 227
Shelford's law of tolerance, 191–92
Shults, F. LeRon, 39, 119, 127, 137–38
Sigmund, Karl, 61
Simmons, I. G., 71
Simon, Julian, 242
simple living, 214
sin, ecological, 138–44, 229
  as attempt to replace God, 139–40
  as refusal to accept our finitude, 140–41
  as unwillingness to face reality, 141–42
  ecological consequences of, 142–43
  holistic nature of, 143–44
Sittler, Joseph, 47, 63, 64, 81, 86, 98
Sleeth, Matthew, 234
Sleeth, Nancey, 246–47
Smith, Christian, 66, 70, 74
Smith, Stephen, 250
snowshoe hare, 158–59
sociality, 59, 65, 230
Solomon, Steven, 10, 58, 189
Soper, Kate, 215
Sorrell, Roger D., 57, 150
soul, 14, 33–34
Soulé, Michael, 22
Southern California, 161, 169
Southern Ocean, 169–70
Southgate, Christopher, 72–73
speech. *See* communication.
Speth, James, 204
spirituality, 2
  physical, 2, 12–13, 56, 115, 228
spotted owl, 204
Staats, Reinhart, 214

Stassen, Glen, 29, 147, 211, 219
stewardship, 131–32
stochastic, 257
Stone, Lawson G., 99
Stott, John, 223
Strier, Karen B., 62
Stuart, Douglas, 80, 81
Stuhlmacher, Peter, 148, 224
succession, 258
Sukhdev, Pavan, 176

Tarlock, A. Dan, 8, 195
Taylor, Charles, 41, 49
Taylor, Paul, 251
Teale, Edwin, 222
technology, 19, 30–31, 47–49, 67
theocentrism, theocentric, 76–77, 121, 148–52, 206, 229, 258
theology, 22, 77, 258
thinking. *See* cognition
Thompson, John N., 9
time lag. *See* feedback loops
Toulmin, Stephen, 35, 40, 151
Tower of Babel, 140, 201, 242
Towner, W. Sibley., 98
transcendence, self-transcendence, 66
trichotomism, 15
Trinity, 123
Turner, Monica G., 256

Valerio, Ruth, 18, 218–19
Van Dyke, Fred, 13, 22, 60, 78, 179, 198, 204, 207, 217, 223, 236, 239, 243, 246, 251
Van Houtan, Kyle S., 11
Van Huyssteen, J. Wentzel, 59, 74
Vitousek, Peter M., 8, 163–65
Von Rad, Gerhard, 106, 118, 120, 122–23

Waal, Frans de, 6, 70
Wackernagel, M., 8, 172, 179, 182–184, 218
Walsh, Brian J., 52, 241

water, 10, 50, 58
wealth, 178, 183, 185, 211–13, 219, 243
Westermann, Claus, 105, 113–14, 142
White, Lynn, 17, 20
White, Peter S., 197–98
Whitney, Elspeth, 44, 47
wilderness, 45, 258
Wilkinson, Loren, 22, 40, 80, 89, 109, 115, 214–15
will, willed action. *See* free choice
Williams, Patricia A., 61
Wilson, Edward O., 156
Wirth, Timothy, 239
wondering onlookers, 221–22, 232
worldview, 111–13, 248, 258
Worm, Boris, 9, 193
Wright, Christopher J. H., 82–83, 86, 87, 96, 99, 103–4, 106–7, 220

Zeman, Adam, 15